MAN
THE FOREMOST CREATION

MUHAMMAD FAROOQ SIDDIQI
MUMTAZ JABEEN

BlueRose ONE
Stories Matter

First Published in August 2022

ISBN: 978-93-5628-040-3

BLUEROSE PUBLISHERS

www.BlueRoseONE.com

info@bluerosepublishers.com

+91 8882 898 898

Cover Design:

Aveek

Typographic Design:

Rohit

Distributed by: BlueRose, Amazon, Flipkart

Dedicated to the parents and siblings of the first author for their motivation to understand deeper meanings of the Qur'an

Prologue

The subject matter of this book involves consideration of two varied aspects: One relates to what is already known to mankind, while the other falls into the pre-historic period.

The first one relates to the formation of man in ontology, where every stage of growth of foetus, starting from germ fluid to a fully formed baby, is known. The baby grows into maturity and old age when death overtakes him. The other side of the story of man in its phylogeny in nature, from the formation of the first cell to a full-fledged man, is a matter that defies positive scientific proof and can only be told by the Creator. This book tries to get help from Divine Books to see if a clear picture of the story of life can be obtained.

Since the beginning of life about 3.5 billion years ago, its development has witnessed many anatomical and morphological changes. No one can deny that the original anatomy, physiology and morphology of animals owes its origin purely by the Creator. The evolutionary changes by the long-time impact of local environment and genetic mutations over generations and generations of existence was brought about in the form and function of the body. The basic formation by the Creator, was however improved constantly and new species continued to be created until the final product, the perfect man, came into being.

Biologists have tried to ascertain the sequence of various life forms existing in different time-periods all over the Earth. As the study relates to an extreme remote past, they have found fossils to be better tools to unravel the origin of new life forms over a long period of time. Fossils are formed when hard remains of organisms like skeletons, scales, teeth, shells and bones are buried in the sediments for thousands of years in shallow or deep layers. The age of fossils is determined by radiocarbon dating or by relative dating, i.e. the position of the sediment layer where the fossil was deposited with reference to one and the other. The geological history of the fossil can thus be ascertained. The fossils present in the deep layer, will be older than those deposited in upper strata.

The origin of various life forms is related to the story of man based on fossil records or geological history, it, reveals that unicellular organisms were the first to appear on the Earth, followed by other invertebrate and vertebrate animals. Important invertebrates besides sponges, jellyfish and marine starfish at that point of tim were insects, worms and moss animals. The vertebrates with backbones like fish, amphibians, reptiles and mammals are formed later. The group of mammals includes a large variety of animals like cats, dogs, rodents, elephants and primates. Man belonging to the group of primates like gibbons, gorillas and chimpanzees, is the final product.

Geological history is a good, scientific way of describing the sequence of the origin of various animals, but it totally overlooks the opinion of the Creator. The present work is an attempt to remove this lacuna and give a better picture of the origin of man. The only course left is that, coupled with any scientific proof as to how and when everything was created, one can look into the Divine Books and see what the Creator himself tells us about the formation of various shapes and varieties in His creations.

The history of mankind reveals the presence of many enlightened people over different periods of time, who tried to explain to the masses. In general, the presence of a Supreme being who created them and that all should follow the path of righteousness as suggested by Him. Based on different points of view, there are currently thousands of religions in the world, of which about two-thirds of the believers in God follow Christianity, Islam, Judaism, Hinduism and Buddhism.

Many religions possess coded instructions and the believers respect and worship their holy books.

God has sent Prophets over different periods of time for the guidance of humanity. Some were the Messengers revealing the words of God that were saved in the form of Divine books. Prophet Muhammad, the last Messenger, received the final revelation of God's words preserved in the form of Qur'an. Qur'an mentions four other Books revealed to earlier Prophets, viz. Scrolls/Suhuf of Abraham, Torah/Taurat of Moses, Psalm/Zabur of David, and Gospel /Injeel of Jesus.

The Scrolls of Abraham are now considered altogether lost. The other three Divine Books before Qur'an are not presently available in their original language but are their translations that as such suffer from mistakes or human errors of interpretation. At the time of their revelation the means of preservation of revealed messages were extremely poor. Large parts of these books are not the true words of God. Many of the 'wise men' or followers of the Messengers of these Books have altered the texts to suit their immediate requirement at the cost of originality and truthfulness. These therefore, cannot be considered as a true guide to know how God did create the Universe and all living things including man.

Qur'an on the other hand contains original words of God, well preserved and unaltered by any human interference. It is pure and present in the same language it was revealed. People did not believe on Muhammad but questioned him about the authenticity and truth of what has been revealed (41: 41-42; 32: 3).[1] God Himself had guaranteed its security: "Indeed, it is We who sent down the Qur'an and indeed, We will be its guardian." (15: 9).

---

[1] Figures mentioned within the brackets in the following pages throughout this book refer to the number of Surah (Chapter) followed by the number of Ayah (Verse) in the Divine Book, the Qur'an, revealed to Prophet Muhammad by Almighty God. The Verses quoted in various places in this book are in Arabic and their English translation by recognised translators is used.

In view of its purity and authenticity, it is better to delve deep into Qur'an and find out what the Supreme Creator Himself says about the processes and nature of his creations. Scientists who studied the facts mentioned in Qur'an agree that no one knewabout them during the time it was revealed. God Himself points that out: "We will show them our signs in the horizons and within themselves until it becomes clear to them that it is the truth" (41: 53).

The truthfulness of the Qur'an that it is the word of God was proved when the latest scientific research critically examined various sayings of the Divine Book and found them to be absolutely correct. A few examples selected from the Qur'an relating to the atmosphere, hydrosphere, lithosphere and biosphere will suffice to satisfy any inquisitive mind that they can only come from God and that the Qur'an is the true Divine revelation.

Qur'an reveals a fact aboutAstronomy that no one could have known at the time of Muhammad and even scientists knew it only in the middle of the twentieth century. It is revealed that the planets, sun, galaxies and the whole Universe were once one mass and God separated them with great force (21: 30).

The fact of the Universe's being built with power or with powerful explosion or with a big bang was known only in 1931 when Edwin Le Maitre, an astrophysicist, suggested the beginning of the universe through the explosion of a tiny particle. This tiny extremely dense and hot mass that came into existence by the will of God, exploded with a loud bang and formed the entire space, matter, time and energy. God not only exploded the tiny particle, He forced it to expand constantly (51: 47). The fact that space is continuously expanding was first pointed out by Edwin Hubble in 1929, who observed that all the galaxies were moving away at great speed.

Another important fact relates to the movement of celestial bodies in orbits. Previously, scientists believedthat the sun was static and the earth was moving in an orbit around it. But they have now discovered that even sun, moon and the earth are floating on their own orbits, a fact already mentioned in Qur'an (36:40).

A strange situation in the hydrosphere is mentioned in Qur'an about a place where two free-flowing water bodies do not mix with each other. They do not encroach one upon the other (55: 20). The meeting of two waters cannot be seen by the naked eyes as both sea waters look similar. But modern scientists measuring the salinity, density and temperature at varying depths of Atlantic Ocean and Mediterranean Sea at their meeting point near Gibraltar were astonished to witness the truth of Qur'anic words , "There is a barrier between them", owing to which they do not mix but remain separated with their own individual characteristics. At another place in Qur'an God informs that there are layers of darkness within an unfathomable sea (24: 40). The truth of this information was manifest recently to scientists who, with the help of modern equipment, could measure that no significant light can penetrate oceans below 200 metres.

Turning to lithosphere, We come across an example of mountains and their roots: "Have we not made the earth as a bed, and the mountains as pegs?" (78: 6-7). Mountains appear as a solid mass of stone, sand and dust lying on the earth's surface. Geologists have now confirmed that the mountains too have roots going down below the surface that look like pegs, as revealed by Qur'an. At one place it is stated in one place that mountains standing firm with their roots prevent the Earth from shaking (16: 15).

Another example relates to the breaking of the earth into different continents: "And by the Earth, which breaks out: surely this is a decisive Word.." (86: 12-13). Geologists now agree that the earth once was a single unit called Pangaea, which broke up around 200 million years ago. And, the broken parts according to Alfred Wegener have drifted away and form present day continents. Recent theory of plate tectonics mentions that the earth has broken into 7-8 plates. These rigid plates lie on top of a partially molten material over which they move slowly. In their movement they either converge or diverge in relation to one another.

A biological fact mentioned in Qur'an relates to sexual procreation by a small quantity of mingled fluids (76: 2). The verse reveals two very important facts about semen: One that an infinitely small quantity of semen called nutfa in Arabic, is used in fertilisation, and the other is that it is a mixture of male semen and female ovum that causes fertilisation. Only a small quantity or only one single cell, spermatozoon, out of over 50 million ejaculated by a man during sexual intercourse will actually penetrate one ovule. From this mingled fluid or germ-fluid, He made an offspring (32: 8).

The birth of males or females is determined by the nature of mixed nutfa (53: 45-46). Modern scientists, by studying X and Y chromosomes in both semen and ovum molecules, have found that it is this small quantity of mixed semen that determines the male or female sex of the new-born.

God has mentioned that, like human pairs of male and female, He has created everything, living or non-living, in pairs. Plants too have both sexes, a point known to men only in the last century. As, yet we do not know more things that are created in pairs, according to God (36: 36).

The narration of the stages of development of an embryo in the mother's womb is a clear sign of the Qur'an being the word of God, as men had no knowledge about it until microscope was developed. Anatomists using modern instruments in the 19th and 20th centuries witnessed the truthfulness of Verses in Quran that God creates in a mother's womb in three veils of darkness (39: 6). Seeing the truth of God's words, a famous embryologist and anatomist, Keith L. Moor (2011) accepted that the Quran contains accurate passages about embryonic development.

Besides scientific information, God foretold Muhammad in the Qur'an some events that occurred in his lifetime or that would appear later. It was revealed that the Roman Empire was defeated and that they will soon be victorious within *'bidh'un'*, a short period of three to nine years (30: 2-4). The Persians defeated Romans in 613 and after nine years, in 622, the Romans got a decisive victory over the Persians at Armenia.

The conquest of Mecca and the triumph of the Muslims over the polytheists ware foretold by God to the Prophet. He was given the glad tiding of safely entering,without any fear, Masjid al-haram, with their heads shaved or hair shortened. You do not know that the entry into al-Haram has been arranged before a conquest near at hand (48: 27). Muslims at the time of this revelation were passing through difficult times under unfavourable conditions, and no one could have even guessed that they would gain victory over Mecca.

The case of Abu Lahab, uncle and ardent enemy of the Prophet Muhammad, was an eye opener to the people of the time. It was revealed to him that the hands (sons) of Abu Lahab, his wealth, and all he had gained would perish (111: 1-2). Abu Lahab died, after the battle of Badr, of an infectious disease, and his sons, his powers and his immense wealth were of no help to him. He was pushed by sticks to a grave and buried there,lest his disease infected others.

It was also prophesied that the body of Pharaoh would be preserved as a sign to future generations (10: 92). The truth of God's words as mentioned in the Qur'an and uttered at the time Pharaoh was drowned in pursuit of Moses and his followers, Israelites 3000 years ago, was proved when the body of the drowned Pharaoh was found and kept in Egyptian Museum at Cairo. Maurice Bucaille, who with other doctors examined the Mummy of Pharaoh Merneptah, son of Rameses II, certified that he died of drowning.

All the above-mentioned examples of natural phenomena and other events provide ample evidence that the Qur'an is the Book of God. Its verses were revealed over a period of about 23 years at different times as the contingency of the day required. The verses referring to one subject in different ways therefore may be found at various places of the Book. The description of historical events is not given at one place but is scattered under different context. One has to look at different places to get the whole picture of what had happened in the past about that particular event. This is not the case ofonly historical events, but all the subjects relating to religion, philosophy, science or others get the same treatment and are described in various places in the Qur'an. Moreover, the narration of a subject may suddenly change to include a similar fact aboutthe other subject in the same sentence. The description is not disjointed but gives two sides of the same coin in a different context. For example, if rising of plants from the dead earth is the subject, the same sentence may mention that man will once again rise after death. Here, rising after death is the common factor of the sentence, though it deals with two different subjects. The relationship can be found if one considers plant growth with water rising through various biological, chemical and physical processes. The resurrection and standing before God will also be subject to various processes including space-time manipulation. More often if the stages of occurrence of a phenomenon are being described, the sequence may be broken and the last stage may be taken up sooner than describing the first. Or only the first and middle steps may be given, avoiding any mention of the final stage. If one has to look for any topic, he will have to collect its gems scattered all over the book.

The beauty of Qur'an is that no one will find any discrepancy, disparity or contradictions in the description of various topics, although their narration occurs scattered over many places. This again indicates that it is a Divine Book. It is also a fact that not a word in Qur'an has been changed; it is preserved in the same way it was revealed.

Qur'an is not a book of science but of signs-- the physical and natural phenomena are amply taken care of. The scientific facts are mentioned in 750 Verses (about one-eighth) of the whole Qur'an relating to Pure Sciences, Life Sciences, Technology and Astrophysics.

Infact God exhorts His subjects to ponder over the intricacies of His creations through science and signs and to learn about

His Power and Command. He is the one Who has provided faculties of hearing, seeing and understanding to analyse and to know the functioning of man and all other creations of God (67: 23). Men are asked to move around and see the work of God in order that they may sharpen their wisdom and power of hearing (22: 46). The use of various faculties is further explained in one long verse that exhorts wise men of reason to critically observe the Signs of God presented to serve humanity in the form of heavens, skies and the earth; in the alternation of the night and the day; in the sailing of the ships through the ocean for the profit of mankind; in the rains that make dead earth bubbling with life; in the beasts of all kinds that He has scattered through the earth, and in the moving winds and clouds (2: 164).

Qur'an available today is original, faultless and pure and can be taken as a true guide to know how God created the Universe and how the living things, including man, were formed. God proclaims no less than 87 times in Qur'an that He is the Creator of Heavens and the Earth and whatever life is present in them. God has planned, designed and created everything living and non-living that is in the heavens and the earth.

Contents

Note: Matter within the brackets indicates number of Stages mentioned by God in Qur'an through which life formations have passed until the creation of man, the Final Product.

Introduction

All the creations of God, whether they are celestial bodies, or earthly mountains, rivers, plants, birds, animals, or man, show a remarkable perfection and artistry in their formation. They are found in a variety of size, shape, colour, texture and structure and have close relationships with each other. Considering them together, they belong to two major groups of animate and inanimate objects. However, the two groups are similar because they are formed of the same elements. All atoms have a nucleus around which electrons revolve and they interact with each other to form molecules, and molecules interact to create chemicals. Both living and non-living objects follow similar laws of chemical actions and reactions and are governed by similar physical laws. The inanimate creations are lifeless and dead, they have no power to grow or procreate. On the other hand, 'life', makes the living organisms capable of utilizing outside materials like light, water, gases or food as energy and then using them for their growth and reproduction.

The whole mosaic of creation, whether animate or inanimate, seen or unseen in the universe, works in perfect coordination and, as such points to a Supreme Creator, God, who not only created them but controls them through His pre-determined Universal Laws. All his creations shows a well-designed organization that functions successfully with faultless precision. God has the power to control and accomplish all of his intentions regarding his creations (85: 16).

As far as living matter is concerned, the earth is actually a planet bubbling with life of varied forms, shades, and colour. It is inhabited by a variety of life forms that spread over every type of environment, ranging from the extreme cold of the poles to hot deserts of the tropics; from sparse scattered grasslands of the prairies to dense evergreen forests of the Equator and occupying lithosphere, hydrosphere and atmosphere. Their size ranges from micro-organisms to giant-sized whales and their shape varies from thread-like worms to heavily built rhinos or elephants. They display a remarkable pattern of a variety of hue and colour seen all over the earth.

Plants and animals are almost completely recognised, but hundreds of thousands of insects and a variety of unknown species that live in deep forests remain undiscovered. If mites and nematodes are included, the total number of species on the planet might be in the millions. Scientists have been able to characterise roughly 1.5 million species so far, but it is estimated that there may be a total of 10 million species. They were estimated to be 30-50 million in 1990 by Mc Neely and others (Mc Neely, J.d. et al). (1990).. About 15000 new species are added each year to the list of known species and with this rate for 30 years from the last count in 1990 to 2020 about 4,70,000 (15000x30) may be included in their lowest estimate to make it about 3.5 million of the total species described so far. The number of individuals in each species may run from a few to thousands and thus the total number of organisms inhabiting the earth is exponential.

All the varied formations existing on the earth surface, in the oceans and in the air have been created by God. The Creator Himself has commanded man in Quran to travel over the earth and see how all the living and non-living things were created (29: 20). Present work is a follow-up of the Command to see how He has created the life forms on planet Earth and produced His final creation (the Man) (29: 19-20).

God, the Perfectionist

Observation of all the creations of the Universe points to the involvement of a perfectionist in their creation. They show a remarkable symmetry in the arrangement of the body parts, whether it be the branches, stalks, leaves and flower petals in the vegetations or eyes, nose, fingers, arms and legs of the animals. All have been created in many shapes and structures. The shape of animals is more striking as they have been built in every imaginable form: thread or ribbon- like, star or saucer-like, spherical, oval or cylindrical.

The art of the Creator is displayed in varied colours over every object, like plants and fruits, rocks and minerals and humans and animals (35: 27-28; 16: 13).

The beauty of creation is also evident from the fact that his animate creation repeats itself. All of His creations are in pairs: "(Allah) that had created pairs in all things..." (43: 12). Plants have male and female sexes, which was previously unknown. However, everyone today recognises the fact that plants have their own couples. : "With it (rain) have We produced diverse pairs of plants each separate from the others." (20: 53). Beside the produce of earth and the living organisms, there are also items created in pairs but are invisible to the naked eye. For example, in micro world protons and electrons of atom, black hole and white hole (or quasars) in the Universe and many others which are not yet known to man (36: 36).

The abundance and scarcity or the relative existence of all the creations seen in the world, are not by chance but are the result of the special design of the Creator. He has provided immense reserves of precious rocks and minerals, has provided abundant supply of plant produce, has reserved huge stores of water, solar or atomic energy, and stored treasures of various other items in the Universe. Every reserve, store, or treasure is the property of God and He keeps special watch over it. Only a restricted quantity is released for use according to the requirement in order to keep the general balance in the demand and supply. He maintains due balance in the growth from the earth of all kinds of means of subsistence (15: 19-20; 54: 49); "...And we send it not down save in appointed quantities." (15: 21).

A perfect example is provided by atmospheric gases that are kept in balance through proper management. Their quantities or measures remain constant in spite of their use and circulation in the atmosphere and among the organisms and other elements.

The proportion and balance maintained in the creation of the Universe is precise and faultless. God has created seven heavens one above another and wants observers to see if there is any lacuna in proportion therein. They may try again and again, but their sight will return weakened and they will fail to find any flaw (67: 3-4). The creation of the Universe and the maintenance of order and balance amongst the heavenly bodies, however is a difficult task compared to the creation of man (79: 27).

The heavenly bodies are remarkable for their perfection in an orderly and proportionate arrangement. All of them function in unison without disturbing the mechanism of one another by keeping a precise distance from each other as determined by God (79: 28).

Various objects in the Universe are arranged in precise mathematical measurements between them. The earth is kept at a pre-determined distance from the sun so that it receives only the required amount of solar energy. Revolution of the earth around the sun in one year causing variations of season, and rotation of earth in twenty four hours leading to the formation of day and night have already been set. The occurrence of day and night is essential for activities and rest of living organisms. Had it not been so, it would have been hazardous for life. The importance of this setting is explicitly emphasized by God's questioning if the night is made everlasting till the Day of Resurrection, who is there to bring you light? Similarly, if the day is made everlasting till the Day of Resurrection, who could bring night for the rest (28: 71-72).

All of God's creations share one characteristic: they grow and develop in a well-proportioned and measured manner after creation (87: 2-3).

The growth and function of all the body organs is in proper order and balance, and this is mainly due to laws of performance determined by God. He created water on the planet earth but fixed only a small proportion of it for organisms usage. It is this small percentage that enters the water cycle and is distributed by God in measured amounts throughout the planet according to the needs of each location (43: 11). The systematic function of water cycle runs its course as ordered. The winds raise clouds and spread them along the sky, and cause them to break in the form of rain (30: 48). Order and perfect balance are also witnessed in the transfer of energy from one organism to the other for proper functioning of the ecosystem.

All Things Created for Man

When God started laying the foundations for the creation of man, He also began to create other living organisms and non-living matter as his favour to be at the service of man, His Vicegerent in the earth and the one created Superior amongst all of the creations. His favours are plenty and no one will be able to count or number them (14: 34; 16: 18).

Among His special favours granted to man are creation of earth as a place of settlement and the sky like a ceiling. Other important favours are the formation and perfection of men in beautiful shapes, and the provision of good things (40: 64).

Everything in the Universe, including the earth, was created for the benefit of humanity and to meet their many needs (2: 29; 45: 13). The night is for sleeping, whereas the day is for seeing stuff and doing work (40: 61). The sun, moon, and stars are created to assist man in a variety of ways (16: 12; 14: 33). Clouds transfer water for man's use from one place to the other and side by side provide shade and resting place to the weary traveler (2: 57). Man has always been guided for direction by permanent natural features and has specifically looked towards stars for finding his way during nights (16: 16).

The earth with its fertile plains, mountains, rivers and seas has been made for the service of man. It is pointed out that the earth's surface has been spread out as a resting place and pathways and rivers have been formed to facilitate easy movement and find a way. Fertile, pliable and manageable plains help easy movement and provide means of sustenance to not only man but to all the other living creatures (67: 15). "And We have provided therein means of sustenance for you and for those for whose sustenance you are not responsible." (15: 20).

Earth produces various types of fruits, orchards and leaves and stalks for fodder. The products of palm trees and vines, husked grain and scented herbs, wholesome drink and food, all are obtained from earth (55: 10-12; 36: 33-34; 16: 67).

Water is essential for the survival of life and therefore is required in abundance. Water helps in the production of grains, fruits, nutritive fruit plants used by man as food and plants for use as cattle feed (80: 24-32). Beside satisfying the thirst of everyone, it is used in agricultural operation. It is required to grow vegetation to be used as fodder for cattle, and also for producing items of human consumption like vegetables, corn, olives, date palms, grapes and every kind of fruit (16: 10-11; 14: 32).

The water bodies have been made subservient to man and provide him various items of use like pearls and corals for ornaments, and fish as food. The seas are made subject and help ships for smooth sailing with their buoyancy. and the rivers too have been subject to you (14: 32; 45: 12; 55: 22). Moreover, the seas have been subjected so that you can obtain fresh and tender meat to eat from them, as well as ornaments to wear from them, and see the ships that sail on the waves.(35: 12; 16: 14; 17: 66).

Beside agricultural produce, there are bees, birds, and animals that have been created for the service of man. Honey bees have been made to form honey that has medicinal power to heal many ailments of man. (16: 69). God sent down some delectable items, such as manna and quails, and instructed them to eat of the good things that He had provided (2: 57).Cattle and sheep are used for their wool, meat, milk and other products (16: 5). Cattle also provide a means of transport to men (16: 7). Besides these benefits the possession of cattle is a source of honour and respect for men. When they lead their cattle to pasture in the morning or drive them home in the evening, they feel a sense of pride and beauty in them (16: 6; 16: 8; 43: 12). Cattle (here used analogically to include all types of means of transport like horses, camels, ship, etc) are used for riding too.

Apart from the favours mentioned so far, there are many more apparent and unapparent that have been bestowed but are still unknown to man (31: 20). They may hopefully be discovered in future. The purpose of mentioning the favours was to highlight the superiority of man among all the creations in the Universe and that whatever is created by God in the heavens and the earth.

The above examples in Qur'an are sufficient to hammer the fact that all the objects of the heavens and the earth have been created by God to satisfy the needs of body and soul of man, the Superior one amongst all the creations of God.

God-A Perfect Planner

God's primary goal in arranging the creation of the universe was to make man His Vicegerent on Earth. He raised some of them to a higher level than others in order to assess their suitability for the post they would be assigned (6: 165).

With this objective as the motive, God the supreme creator chalked out a magnificent plan of creating the Universe— a great organization of Creations. Rules were framed for the execution of the plan and the course of action was undertaken under pre-determined parameters. In fact, God is the best of planners and in all things. He is the master mind of all the plannings (3: 54; 13: 42). No one can find any flaw in his designs and functions. Symmetry, proportion, balance and perfection are the hallmarks of his creation.

The whole design, its execution, the processes to follow, the nature of the outcome and every other item and event was recorded in 'Al-Lauh Al-Mahfooz', the Book of Records, the

Book of Actions, or the Mother Book. Every event that has taken place and everything small and great that will occur in future are all recorded in the Great Divinely Protected Book (22: 70; 54: 52-53). Not an animal in the earth, nor a flying creature, nothing is neglected in tQura'an (6: 38). The details recorded in this Divine Book are final and no one except God has the power to change it. Once every minute detail from the start of the creation to its culmination and then resurrection in the Hereafter was recorded, God commanded by saying "kun" (Be) for the objective to be achieved and "fa ya kun" (and it is done) (36: 82; 16: 40; 2: 117: 54: 50). The word "kun" or the command "be" in the eyes of Creator is executed promptly.

The process of creation, however, depends on certain rules determined by the creator that take time to achieve the goal. "Fa" is in fact the time gap that is in-built in the designed programme for the completion of the work. The time involved in matters of creation may be a fraction of a second or may extend over millions of years for attainment of the objective. Anyway, the process of work runs its own course and is completed according to the design formulated in the Book of Actions and rightly so after God's commands. God knows the result of his command and for him it is fait accompli, though the final stage of the plan of re-creation and the formation of the Hereafter is yet to happen. God can, of course, create any object in a fraction of a second, but if he plans to create a thing step by step under a predetermined plan of action following his rules of creation, it will be finalised accordingly without any deviation from the predetermined plan. God never changes his plan of action (35: 43).

Glorious Qur'an, revealed to Prophet Muhammad, is a part of the preserved Tablet, Al-Lauh Al-Mahfouz or the Mother Book (85: 21-22; 56: 77-78). God has sent it down and has vouchsafed to guard its purity (15: 9). It relates to that part of the Great Plan that concerns the life of man and his environment. The rules concerning the creation of life, its performance, death, resurrection and all other things are pre-planned and noted in the Divine Book.

Creation of Life

The Creation as revealed in the Quran envisages broadly the following items:

a) Creation of Heavens and Earth

b) Creation of Life

c) Creation of a well-proportioned, good-looking man, Adam as a superior being amongst all the creations of the Universe; breathing into him his Spirit to make him His Vicegerent on earth; bestowing upon him the power of observation, hearing, reasoning and analysis and the knowledge and understanding of everything; and granting him power of judgement.

d) Sending prophets and messengers to every community and tribe for guidance to accept the supremacy of God and to lead them to follow the straight path, the only path that is pleasant to him.

e) Making earthly life a test case for humankind who after receiving the guidance from God remain grateful to him and perform acts of righteousness or are not grateful.

f) Winding up of the whole organization of creation.

g) Raising up mankind and other life forms once again in the hereafter for final judgement of men and sending each of them into Heaven or Hell as their permanent abode on the basis of their deeds and faith during their lifetime.

Creation of Man in Seven Stages

Among all the creations of God, the man was created in a beautiful shape, well-proportioned, and perfectly built. Even this supreme creation had to pass through diverse stages of formation until a final good-looking perfect figure was obtained (71: 1; 7: 11). God, while describing the formation of man in stages, has pointed out that it was like the growth of plants from the earth. The specific use of the word 'rabata' or growing of plants or trees, confirms the similarity in the formation of man and the plants. The man was created from earth and had to pass through different stages of growth until a full-fledged human being was formed. It is like germination of seed in the earth, formation of shoots, and growth of a full, mature plant. In another place, it is highlighted that God started the creation of man, fashioned him into a proportionate functionary, and breathed into him some part of His knowledge and sense (32: 7, 9).

God has mentioned that man had to pass through various stages of formation from the day he started to be created from clay to his final winding up in death and resurrection in the hereafter. While describing the stages, God clearly points out that there are seven pathways or stages through which life forms had to travel and that He had always been watchful about the plan of creation (23: 17).

It is possible to identify these seven stages if one considers Verses 12-16 of Chapter 23, Al- Muminun together with Verse 5 of Chapter 22, Al-Hajj. The identification is easy to count because six special conjunctive words '*tsumma*' (then afterwards)[2] differentiate the seven stages from each other.

---

[2] At one or the other place, all the verses referred to above use the Arabic word 'tsumma' translated as "then afterwards" or "moreover". It is a conjunctive word that denotes i) sequence of events, ii) continuity between sets of different characters, and iii) time involved in the change.

When the word Tsumma is used during the formation of a child in the womb, the sequence, continuity, or change -over may be short and may occur in hours, days or months. But it may spread over thousands and millions of years when the events of creative development, in general, are taking place that a long time for any physiological and morphological change. The word, therefore, acquires special meaning when matters relating to creation in general are being described. Under such conditions, if simply translated as 'then', 'afterwards' or 'moreover', its appeal is weak. The author has therefore translated it in the above verses as 'then afterwards' to be more forceful, carrying its true spirit and to be identified as a special word.

These seven stages are:

Stage 1. "Verily We created man from an extract of wet clay" (Ar. *Wa laqad khalaqnal insaana min sulalatim min teen*) (23: 12);

Stage 2. "Then afterwards We placed him as nutfa (a drop of germ-fluids) in a safe lodging..." (Ar. *Tsumma ja'alnahu nutfatan fee qararim makeen*) (23: 13);

Stage 3. "Then afterwards We made the nutfa into an alaqa (a leech-like clinging entity..." (Ar. *Tsumma khalaqtan nutfatin alaqa*) (23: 14);

Stage 4. "...Then afterwards (We created you) out of mudgha (a lump of flesh), partly formed and partly unformed. (Ar. *Tsumma mim mudghatim mukhallatin wa ghairi mukhallatin*) (22: 5).

Stage 5. "... then We made out of that lump of flesh bones and clothed the bones with flesh (muscle); then afterwards We developed him into final creation...! (Ar. *fa khalaqnal mudghata izaman fa kasaunalizama lahman tsumma ansha'nahu khalqan a'khara*) (23: 14).

Stage 6. "Then afterwards, you surely die." (Ar.*Tsumma innakum ba'ada zalika lamaiyetoon*) (23: 15).

Stage 7. "Then afterwards indeed, on the Day of Resurrection you are raised (again)." (Ar. *Tsumma innakum yaumal qiyamati tuba'soon*) (23: 16).[3]

---

[3] Each stage of human formation is recognized by a special Arabic word that represents the main distinguishing character of the group. The names used in this way are: 1. Teen; 2. Nutfa; 3. Alaqa; 4. Mudgha; 5. Ma'khar; 6. Maiyetoon; and 7. Tub'asoon. Out of these, three words, i.e., nutfa, alaqa and mudgha, have been derived from terminology used for various stages of development in an embryo. These also show the chief characteristic features of animals that were created during different stages of their creation. The English translation of the words given above fails in many cases to carry correct meaning of the Arabic word. The nature of the word and its correct meaning shall be explained, if required in the following pages when each stage is considered in detail.

The system of classification is purely a processbased specifically on the internal structure or anatomy of the organisms. The successive changes in the formation of internal structures are considered in such a classification and therefore bring out a phylogenic relationship between them. As the subject matter dealt with by God exclusively relates to creation of life-forms leading finally to creation of man, the classification considers only animal.

Classification by God separates successive groups on the basis of their natural characteristics. The three characteristics of birth (first from clay and then from nutfa) to death and resurrection are common to all the creatures, irrespective of their differences in morphology, anatomy, cytology, physiology and behaviour. The other groups of creatures are arranged according to the affinity in differentiating character introduced during successive stages of their formation.

Of these seven stages of creation in phylogeny, four also occur in the womb of mother: "If you have a doubt about the Resurrection, (consider) that we created you out of dust, i) "then afterwards out of germ fluid, ii) "then afterwards out of a leech-like clinging entity, iii) "then afterwards out of a lump of flesh, partly formed and partly unformed iv) "then afterwards do We bring you out as infants..." (22: 5).

The point worth emphasizing is that identical words/names are used to identify pathways of formations in the wombs (ontogeny) as used for creation in nature in general (phylogeny). These common words of identification in both situations are: i) Nutfa, ii) Alaqa, and iii) Mudgha. The fourth stage in the womb is birth which in this case is denoted by the word 'tifl' or a child or a baby. In phylogeny however, this stage is referred to differently as 'Ma'akhar', or the final creation/birth of a final creature (man). The use of similar words for stages in the two different but identical situations, make it difficult to understand whether the creation in Qur'an is about one or the other. The matter assumes greater complexity by the fact that Qur'an being not a systematic book of science, the description of a phenomenon is found scattered in various places justifying the point being considered according to the situation. Moreover, all the stages of creation are not always described at one place. God will mention one or two stages to highlight a point and the rest can be understood by inquisitive mind. For example, the first two of five Verses revealed to Prophet Muhammad asking him to read in the name of the Lord. And that he had created man from a clot of blood are most significant (96: 1-2).

Man is said to have been created from alaq or a clinging entity. In view of the importance of the first revelation, the fact mentioned here should be an important beginning of creation of man in general. It can however also be inferred to mean the formation of man from alaq in the womb. Both explanations can fulfill the purpose of the seeker of truth when God mentioned to have created you from clay, then out of a drop of germ-fluids, and then fashioned you (18: 37). As the verse refers to a lowly status of nutfa, it clearly points to the formation of man in the womb and not in general.

Similarities in the successive formations of life in general and man in the womb were first noticed by an eminent German biologist Ernst Haeckel. Enthralled with this discovery, Haeckel propounded in 1866 his Biogenetic Law popularly known as his Recapitulation Theory. This theory envisages that animals in their life starting from fertilization to birth in ontology repeats or recapitulates in condensed form during evolution in general. The theory was briefly put forward in three words: Ontogeny repeats phylogeny. Ontogeny is the life history of an animal, while phylogeny is the evolutionary history of the race of the animal. Ontogeny repeats phylogeny means that an organism repeats its ancestral history during its own formation.

As an explanation of his biogenetic law, Haeckel compared the embryonic development of various species and pointed out the similarities at every stage of their development. The Fertilized eggs of most animals are similar because life begins with a unicellular organism. Later development of embryos of a fish, a frog, a turtle, a bird, and a man resemble each other so closely that it becomes difficult to identify each other. Furthermore, an embryo of a mammal passes successively through fish like, amphibian-like, reptile-like, and bird-like stages during its growth. It was thus proved that the embryonic development of an organism (ontogeny) is a repetition or recapitulation of some of the phylogenic or evolutionary steps. It may be said that man during embryonic development shows the anatomical characteristics progressively of a fish, an amphibian, a reptile, and a man.

Haeckel was an evolutionist, and his whole thinking was oriented in the light of ideas proposed by Darwin. The theory of natural selection propounded by Darwin declares that owing to an increase in population pressure, the organisms struggle for survival in the presence of limited resources. While doing so, only those fit to fight will survive. The generations of these fit individuals have survived by developing special traits to protect themselves from the vagaries of the environment. These useful traits are passed on to new generations. With every generation, the organisms develop some new variations and thereby adapt to new environments in their struggle for survival. With the passage of a number of generations, the adapted variations accumulate and new species evolve. In Darwin's view, evolution is a gradual process and new species evolve from old with the passage of time. On a large scale a phylum gradually evolves and over a period of time gives rise to a totally different new lineage.

Thinking along the same lines of continued evolution and trying to remove various disparities in the specimens, Haeckel faked his drawings of different embryos to prove their identical evolution. The modified drawings of embryos appeared similar to one another. His ideas were acclaimed widely and occupied the minds of biologists for almost a century. The forgery, however, was bound to be discovered and by the middle of last century, the biogenetic law was discarded.

Haeckel (1866) presented a new "Theory of Recapitulation" that animal during its life, starting from fertilization to birth, repeats or recapitulates in condensed form its early evolutionary stages. He may have been greatly elated had he known how God did create organisms and that his theory of "ontogeny repeats phylogeny" was correct. He need not have faked the classification had he seen what was revealed in the Qur'an by none other than God, the Creator Himself. He may have compared embryos of various creatures of all stages pointed out by God and seen that each of them tallies in general aspects of their structure and function.

PART-I

Life in Water

(Stages I to IV)

CHAPTER I

Origin of Life

Living organisms are primarily formed of inorganic compounds of water and minerals.

Water is essential for the survival of every living being and its necessity as a medium of creation of life is pointed out by God: "And Allah has created every animal from water..." (24: 45); "…We made every living thing from water…" (21: 30).

As far as minerals are concerned, scientists have so far recognised 109 inorganic elements of which 25 are essential for living organisms. However, of these only 6 are major elements that form 98 percent bodyweight of each organism whether it be small bacteria or a complete man. These essential elements are oxygen, carbon, nitrogen, hydrogen, sulphur, and phosphorus. The remaining 2 percent of the bodyweight is made by 4 minor elements like calcium, potassium, sodium, and magnesium, and 15 trace elements.

Earth soil or clay is the basic source of minerals involved in the formation of living organisms. The nature and kind of earth (soil) or clay used in the creation of the first living matter is described by God in many ways. In many places in Quran dry earth 'turab' or dust is named as the matter of creation (18: 37; 22: 5; 30: 20; 35: 11; 40: 67). Finer particles of earth soil when wet make clay that has also been referred to in the formation of the first living cell (6: 2; 38: 71).

A different state of soil described in the Quran indicates that it was moulded mud or potter's clay from which the first single-cell was created:(55: 14; 15: 26). It is further elaborated that man was created from *sulalatim min teen* or an extract of clay: "And indeed, We created mankind from an extract of clay." (23: 12).

Creation from potter's clay indirectly points to the presence of heat required for baking the moulded pottery. In a nutshell, it is emphasized that the first living cell was created from clay in the presence of water and heat. Clay and water provided essential inorganic raw materials for the formation of life. It also required a suitable atmosphere to help achieve proper interaction of these materials to be able to create life. As a general scheme of creation clay and water or the earth and water bodies together with their atmosphere were the first to be created by God.

Materials for Life Formation

Land and water, basic materials for the creation of life, actually came into being, according to God, from nothing (19: 67). When God planned to create man, He started his work by first creating from nothing a nano-particle called by scientists 'a God Particle'. This dense matter concentrated at a single point was exploded or blasted by God with great force.

According to modern-day calculations, the tiny dense matter exploded about 13.82 billion years ago with a mighty force and with a mighty sound known today as Big Bang. The explosion not only led to the formation of the Universe, but the whole matter was made to expand which it still does: "We have built the heaven with might, and We it is Who make the vast extent." (51: 47). The expanding matter ultimately produced most of the atoms of hydrogen, helium, and traces of lithium. Giant clouds of this matter, later on, coalesced through gravity and formed stars and galaxies. God informs about these cosmic formations that the heavens and the earth were a joined entity, and that He separated them..." (21: 30). It is generally believed that our star, the sun, and its system of planets, moons, and comets likewise was a cloud of dust and gas or a Nebula around 6 billion years ago. It consisted of 99 percent hydrogen and helium and of only 1 percent heavier and metallic elements. This rotating dense cloud some 3 million years later began to collapse and contract. The collapsed mass accumulated at the centre and formed our sun.

At this point in time, it is believed that another greater and more powerful Nebula passed close enough to pull with its greater gravitational attraction an arm of the mass the of sun. The larger Nebula slowly passed away losing its gravitational force and left behind the extended mass as an arm. This extended mass too like its parent mass began to condense and in the process was broken into smaller parts, giving rise to planets. The planets nearer to the sun were small and in due course of time cooled and condensed into earth-like form, while the planets formed away from the sun were larger in size and remained gaseous.

Another opinion regards that a major chunk of the dust particles contained in the cosmic cloud or Nebula condensed and formed a large and extremely hot Sun while the remaining particles of the scattered dust that were revolving around the sun condensed into small clusters of planetesimals. The planetesimals gradually increased in size owing to the accumulation of more condensed dust. Sometimes the planetesimals themselves collided with each other and made a larger body. The larger planetesimals could pull smaller ones by their sheer force of gravity and began to further increase in size. The larger ones even pulled comets that merged with them and made them still large. This way millions of years later the planetesimals and other materials gave rise to nine planets of the solar system that we know of today.

The chunk broken from the main cloud of dust and gas known as Earth came into being around 4800 million years ago. The hot, spinning ball in the initial stage was formed of gases and vapours of elements. Each element was in the atomic state as the high heat of 4000-8000°C of the time prevented the bonds holding atoms together from forming molecules. As the conditions cooled down the free atoms of different elements began to come together and metals started to form. The earth began to take shape as a separate entity.

For about 200-300 million years the newly formed solidifying molten earth and the volatile atmosphere was bombarded by meteorites that further added material to it. When the earth cooled further to a semi-molten state it was enveloped by an atmosphere of light elements that still existed in the atomic state. The common element in the early atmosphere was hydrogen, but elements of nitrogen and carbon were also present in abundance.

The atmosphere had no free oxygen because any amount that could form from the reaction of ultraviolet rays and carbon dioxide could not live long or was lost. The gravity of the earth at that time was low. Some of the atmospheric elements were escaping into the space as low gravity of the earth could not hold them back. The earth and atmosphere had no hold over each other. God then commanded the smoky heaven and the earth to come close to each other and both agreed to help each other (41: 11).

It was the time when conditions of the atmosphere and earth became suitable for the exchange of gases and earth elements for the benefit of living organisms.

Elements of lithosphere and atmosphere began to come close together at random and produced simple and compound molecules. In the lithosphere metals combined to form metal nitrides and metal carbides. With further cooling, the elements began to stratify according to their weight and density. Heavy metals like nickel and iron (NiFe) sank and settled in the central part and formed the core of the earth. The lighter elements like silicates and aluminium (Sial) cooled at the top and formed the crust of lighter material. The moderately heavy metals like magnesium and some lighter elements like silicates (Sima) got stratified into solid rocks in between the crust and the core. All these deposits also helped in increasing the gravity of the earth.

The surface of the earth cooled down into a solid crust. The solid crust got differentiated into two types according to the thickness and chemical composition. A thick layer composed mostly of silicon dioxide, being slightly light, formed elevated continental mass. The other comparatively thin layer composed of calcium, iron, and magnesium, being slightly heavier, made low-lying portions of the crust.

Contraction of the earth due to its gradual cooling fractured and folded the earth's surface, caused large blocks of the earth to move and form mountains and ocean basins, and resulted in volcanic activity that fractured and moulded the earth's surface. Large and firm mountains with deep roots were set on plains to prevent shaking and to maintain balance (13: 3; 15: 19; 31: 10).

Though the earth has cooled sufficiently to make a solid crust, still about 3800 million years ago it was a burning mass. It was a violent time on our planet. The thin outer shell was still steamy hot. Volcanoes spewed ash, fire, and dust into the air. The surface was shrouded by thick clouds of dust and ash.

Elements of the early atmosphere produced molecules of hydro, nitrogen, carbon dioxide, ammonia, methane, and water vapour. Free oxygen was not available as whatever free oxygen was formed it soon combined with hydrogen atoms and converted into water vapour (H_2 O). Besides water, hydrogen donated its electrons to other substances and formed ammonia (NH_3) and methane (CH_4). With the donation of electrons, the main atoms of hydrogen got reduced. The reducing atmosphere with the absence of free oxygen was called the primary atmosphere.

The intensity of the reducing atmosphere gradually slowed down and a mildly reducing atmosphere was formed. It consisted of water vapour, carbon dioxide, carbon monoxide, nitrogen, and small amounts of other gases. Still, there was no free oxygen. It was present as a combination in water or as oxides. However, the negligible amount of oxygen was formed as it was released from the water after its reaction with sunlight. This oxygen too did not live long as it soon reacted with iron or sulphur compounds.

Free atoms of hydrogen and oxygen had formed water but under high temperature it remained as water vapour in the atmosphere. This water vapour surrounded the earth like an opaque curtain. The curtain obstructed sunlight and the temperature intensity on the earth surface was reduced. When the temperature fell below 100°C condensation of vapour took place and hot water started falling on to the surface of the earth. However, owing to still hot surface of the earth, the water immediately evaporated and went back to the atmosphere. It condensed to form water droplets and fell on the surface of the earth as rain. Such a process of heat, vapour, thunder, rain and then again vapour and rain continued for thousands of years. The process of gradual reduction of temperature explosive atmospheric conditions ultimately came to an end, conditions stabilized and the water cycle started to function. Heavy rains continuing for thousands of years filled depressions on the earth's surface and formed lakes, seas and oceans.

The materials required for the formation of life, i.e., inorganic minerals and water, were made available and the time was ripe for the beginning of Life.

The mention of water and clay as the material for cell formation is not sufficient to explain how life was injected into this dead material. Many scientists and philosophers had been busy in trying to find out the nature of the origin of life. Some early scientists as late as up to middle of 17th century proposed the theory of Spontaneous Generation or Auto biogenesis or Abiogenesis whereby living beings were thought to have been formed from non-living material spontaneously. Aristotle (384-322 BC) in his work entitled Historia Animalium (i.e., Investigation of Living Things) is of the opinion that fireflies were born from the morning dew and rotting manure, tapeworms formed from excreta of animals, and crabs and salamanders originated from wet earth and slime. Jean-Baptiste van Helmont (1577-1644) proposed that mice would be born if wheat covered by a dirty shirt soaked in human sweat is kept in a dark cupboard for 21 days. William Harvey (1578-1657) who is known for his observations that life comes from egg (ex *ova omni*). Auto biogenesis of organisms dominated the thinking of scholars for long period and even now some people think that beetles grow from dung, red velvet mites or rain bugs are born from the first rain of the season and stored water gives rise to mosquitoes.

The theory of spontaneous generation of living things from non-living matter was however proved wrong by experiments carried out by Fransesco Redi (1668), Abbott L. Spallanzani (1765), and Louis Pasteur (1862). Their experiments gave rise to Biogenesis Theory that living things are not created spontaneously but life comes from pre-existing life or '*omne vivum vivo*'. Redi placed flesh/meat in three jars, one covered with parchment, the other with muslin and the third remained uncovered. The flesh decayed in all jars, but maggots were born in only one jar that was left uncovered. He concluded that flies entered the uncovered jar, laid eggs on the flesh kept inside and then maggots were born not from non-living matter but from eggs. Spallanzani experimented by boiling nutrition broth in some loosely corked flasks and in some tightly sealed flask. After few days thick growth of microbes was seen in loosely corked flask, while no organisms were seen in airtight flasks. It proved that the air containing microbes entered the loosely closed flasks and contaminated the broth inside.

The final proof of life-generating only from life came from the experiments conducted by Louis Pasteur. He experimented by boiling froth, but unlike Spallanzani's corked flasks, Louis Pasteur took a flask connected to a swan-necked or s-shaped glass pipe. The boiling broth was connected to air through bent neck, but no micro- organisms were noticed even after several days. This was because the dirt carrying microorganisms was deposited at the base of the bent neck. When the neck was broken, the broth suddenly got flooded with micro genitors indicating that the organisms had come from air. These experiments were able to show that life originates from life, but failed to show how life initially originated.

Finding no way of how life originated on the earth, scientists out of sheer desperation began to think that it did not rise here but perhaps came from other planet. Hermann Ebehard Richter (1865) was the first to suggest the meteorite idea that was supported by Arrhenius (1908) and others too. This theory of panspermia or spore theory collectively known as interplanetary theory proposed that life in the form of resistant spores of simple organisms present in some heavenly bodies were carried to Earth with the help of meteors and cosmic dust.

The presence of amino-acids, an indication of presence of living things, found in Murchism meteorite of Australia supports the view of extra-planetary origin of life. Qur'an, the Holy Book also points that God has scattered living organisms all through heavens and the earth (42: 29). The critics of the theory however point out that no organism can survive long interplanetary travel owing to high radiation in interstellar space. Modern space research is of the view that bacteria can survive the adverse conditions of space travel as they were found to be not affected in moon's waterless and airless environment.

One important item however that forcefully negates interplanetary theory is the water that is essential for the survival of organisms. All the inter-space probes of modern cosmic research are unable to prove the existence of water bodies anywhere in the universe. The studies do however point to the presence of water in traces inside the other planets. There are evidences of presence of running water on the surfaces but they are now completely dry. Even if water was present there it was not in suitable quantities or in such a form that it may have contributed to the origin of initial living matter.

Recent studies have revealed many earth-like planets in other Solar Systems of the Universe. Quran reveals that God had created seven Heavens and earths of similar number (65: 12).

But when water, temperature and suitable atmosphere were available here, why one should not believe that life began on our own earth. God too mentions in Quran the fact that He has made everything existing in the heavens and the earth to serve man and that He has created mankind from clay (31: 20; 32: 7).

The first worthwhile effort in this regard was made in the early decades of the 20th century by a Russian chemist Alexander I. Oparin (1924) and a British biologist John B. S. Haldane (1929) who working independently came to similar conclusion. In view of the similarity of ideas the theory was known as the Oparin-Haldane Hypothesis or the theory of Chemical Evolution. They suggested that the primitive organic molecules interacted with one another and formed spontaneously simple organic compounds like sugar, fatty acids, glycerol, amino-acids and organic bases (pyrimidines and purines). In their opinion the following conditions available in those days helped the formation of organic compounds:

1. Atmospheric gases like hydrogen, nitrogen, ammonia, methane, carbon dioxide, and water vapour indicate the presence of primitive organic molecules,

2. Energy required for enhancing molecular reactions was available from: i) high surface temperatures of the earth being more than 60o Celsius, ii) ultraviolet radiations, iii) cosmic rays, and lightning.

3. A reducing atmosphere in the absence of free oxygen that preserved these early formations of compounds. Free oxygen may have destroyed them by oxidation.

4. Lack of life helped in the preservation of early compounds as any living matter would have quickly used them.

According to them, these organic compounds were formed partly in the primitive atmosphere but largely in the oceans as ocean water by that time had an abundance of inorganic molecules brought down in solution by rivers and deposited therein. Some parts of shallow oceans were converted into a 'dilute hot soup' or 'broth' containing many interacting organic molecules. The heat and salts were further provided to such a soup if there was any undersea volcanic eruption. Oparin and Haldane's theory however remained only a theory and they were not able to demonstrate the actual formation of organic compounds by any experiment.

Spontaneous generation of molecules was successfully attempted by Stanley Miller and Harold C. Urey in 1953. They designed an apparatus for simulating the conditions of the earth and demonstrated how organic matter could be formed through abiogenic process. The apparatus contained a glass flask filled with gases like hydrogen, ammonia, methane, and water vapour simulating conditions of reducing atmosphere. The flask also contained electrodes to generate electric sparks of 75,000 volts to simulate energy as that of lightning. There was another flask for boiling water simulating evaporation and circulation and a condenser simulating rain and Haldane's 'soup'. The mixture of gases was exposed to electric discharges followed by condensation and then boiling again. This process continued for 18 days and the boiling and condensation was repeated several times. The product thus formed was found to contain some simple organic compounds that included amino acids like glycine, alanine and aspartic acid, adenine and simple sugars such as ribose. This experiment supported the views of Oparin and Haldane as to how the organic compounds were formed initially in nature. The only difference in experimental formation and natural formation lies in the fact that nature took millions of years for the formation of organic matter that the experiment was created in only a few days. The Miller-Urey experiment was acclaimed as a path-breaking event and occupied the minds of scientists for many years. Later experiments however demonstrated its weaknesses and everyone began to say that there is no way of life coming into existence by chance.

The weakness of Miller's experiment and the hurdles in the way of creation of life were many:

1. The positive results obtained by experimenting with methane and ammonia were questioned. Scientists have found that the primitive 'soup' suitable for the formation of life existed when the atmosphere consisted of water vapour, nitrogen, and carbon dioxide. These gases, if experimented with, never yielded the formation of amino acids. Miller too accepted the drawback that his experiment was not based on real atmospheric similarities.

2. It was presumed that the atmosphere of the time was totally oxygen-free. In the presence of oxygen, any amino acids formed would have been quickly destroyed. Miller had used a trap to save the newly formed organic matter from the vagaries of the atmosphere. Scientists claim that there was ample oxygen in the atmosphere and any amino acids formed naturally without any protection would have never survived.

3. A greater improbability of the formation of life as a chance happening in nature beside amino acids is the formation of protein, and nucleic acids, the main bodies of life. The formation of protein and nucleic acid from amino acids is possible if water molecule is removed from amino acids and nucleotides through the. process of polymerization. As the amino acids were formed in the primordial ocean, no polymerization can take place in the presence of water. As an alternative, it was suggested that amino acids were formed in water and were later thrown onto the land for polymerization. It is a far-fetched possibility or is impossible. But the state of matter requires the presence of water and a dry land alternating with each other for the formation and stability of proteins. Probably the primary amino acids were created in the 'soup of the ocean' and then dried on the surface for polymerization.

As far as the part played by dry land is concerned many studies have found clay to be the most appropriate quick polymerization material. Sidney, W. Fox (1965) and Cairns-Smith (1986) are of the view that clay was the most suitable material for polymerization. Fox suggested that if some rain occurs at right time on the warm dry land along the edge of a volcano, it may produce billion tons of polymers. This idea however was questioned by Miller and Orgel (1973) for the fact that whenever lava solidifies the surface will not be warm enough for polymerisation.

In order to further elaborate on the idea of 'hot organic soup' a German chemist Gunter Wachtershauser (1990) presented his 'iron-sulphur world theory'. The first organisms perhaps originated around hydrothermal vents near seafloor. This energy source may have been easily available through carbon dioxide, or carbon monoxide brought together with hydrogen on the surface of iron pyrites available in the volcanic vent.

The theories of the chemical origin of living matter seem plausible but no one has so far been able to produce any living cell in the laboratory. Scientists have so far found the sequence of protein formation, but it is still a matter of quest to find out how it all started. God has kept it a mystery and no scientist or philosopher has been able to solve the puzzle.

On getting no response from scientists it may be better to look into religion and mythology that played an important role in developing the thought process of man regarding the origin of life. Early thinkers saw the hand of God in creating every living and non-living matter. The theory of special creation emphasizes the creation of everything out of nothing. Everything was created suddenly and independently of each other. This spontaneous creation has not undergone any evolutionary change and remains the same in form, and structure all through its existence. According to Christianity, God created all the living and non-living things in six days (or periods): Materia Prima heaven and earth on the first day; sky and water on the second day; land and land plants on the third day; sun, moon and the stars on the fourth day; fish and fowls (birds) on the fifth day; and on the sixth-day land animals. Adam, the first man was created from clay, and his partner, Eve was created from the sixth rib of Adam.

Islamic holy book the Qur'an reveals that God created man from clay. Accordingly, early interpreters formed a concept that God first prepared a clay model of a man and then said 'kun' (be) 'fa ya kun' (and then it was). There still are some protagonists who subscribe to this theory. There is no mention in Qur'an about the creation of Eve, but taking the hint from Christianity Islamic thinkers too point out that Eve was created from the rib of Adam.

The latest scientific assessments suggest the origin of man from clay by a Super Power. The idea was first revealed in Quran that God created a single living cell from clay and from it created its mate of like (or similar) nature (4: 1; 7: 189). It is from one being (a single living cell) that He, after going through various formations, created a perfect man (39: 6).

Life Formation: Difficulty Level

All philosophers, thinkers, scientists and men of learning from old to modern days of satellites and space travel acknowledge the fact that there is no proof of proteins originating in nature. Its creation is beset with innumerable 'ifs' and 'buts' and to human mind is in the realm of improbability. The creation of life from inorganic elements like minerals or earth and water is a difficult task. The difficulty level of life emerging from the raw materials is comparable in the words of a British astronomer Fred Hoyle (*Nature*, Vol. 294, 1981, p. 105) to the chance that "a tornado sweeping through a junkyard might assemble a Boeing 747 from the materials therein."

Scientists now acknowledge the creation of life to be a difficult task or rather impossible to achieve. The Creator had Himself acknowledged the complexity of the problem and the difficulties involved in solving it. How beautifully He explains the difficulties in the creation of man by comparing it with still greater difficulty in the creation of heavens and the earth. Most men do not understand that the creation of the heavens and the earth is a greater matter than the creation of human beings. He inquires whether men are difficult to create or the heavens that He built (4: 57; 79: 27). The comparison does not lessen the difficulties involved in the creation of man but was to emphasise that if a more difficult task was achieved, the formation of man could also be successfully overcome.

Life Formation: Indications in Qur'an

The impossibility level of creating a single protein molecule becomes still greater if one thinks of making the proteins and nucleotides grow, replicate and become alive. No laboratory experiment will be able to decode the process through which the first building block of life, a functional cell was created. All are forced to accept that some intelligent Super Power is behind every creation. They are not only people of religion that believe in the existence of a Super Power, but modern enlightened scientists do also believe in the powerful role of the Creator in controlling the movement of heavenly bodies, oceans, and all the living and non-living things on the earth.

Scientific studies now accept the basic fact revealed by God that proteins are formed from dead organic elements of water and clay (21: 30; 23: 12). God points out that man, a living organism, was created from lifeless substances (2: 28).

Besides the formation from water and clay, there are also indications in the Qur'an as to how the process of life formation was achieved. In the first instance, conditions were made favourable for the appearance of life by stabilising the lithosphere and atmosphere and ensuring suitable presence of clay and water. This was the time when the great event of the creation of first living matter was to take place.

God had established Himself on a permanent Throne (*Arsh),* a place of Authority or His centre of activity far away in the Heaven. It is from here; he directs the ordinance unto the earth (32: 5).

However, considering the importance and uniqueness of the occasion of starting His mission of creating man from water and clay, God shifted His Throne upon water (11: 7).

Concentration of God at the Centre of Creativity, at this

moment of time is a clear indication that some unusual and difficult problem was to be tackled and that too in water. God had to be present at the centre of activity or where the first brick of life was to be constructed. He had to personally involve Himself in the two most difficult tasks of creation of heavens and the creation man. No angel was given this responsibility. It was achieved by God the Magnificent, all Powerful and Supreme Creator: "Say: "It is He Who has created you..." (57: 23).

God and God alone could have created these complicated items that in view of intelligentsia were impossible tasks. He has the power to do anything, create anything and achieve anything in a fraction of a second. Then why He took a circuitous, time consuming, hindrance ridden route to achieve His goal. He could have directly constructed the clay model of the end product, a man, as many of the enlightened men earlier had thought was actually done, and said 'kun' (Be) and the man would have come alive with all his organs functioning perfectly. Before starting creation, however, God had chalked out a plan of action, determined certain rules and parameters for its fulfillment and then ordered 'kun' (Be) and the Universe and all the things began to work according to His master plan.

God has ordained Laws of motion and performance that must follow (87: 3). Every non-living item in space from an atom to large quasars and every living organism from a bacterium to a complicated object of creation, the man, were created and function according to rules and plan of work. God never changes the rules He Himself had framed and His Word does find its fulfilment in truth and in justice (6: 115). When the rule for the creation of man was recorded that he will be created first as 'nafsun wahid' (unicellular organism) from clay and water, will grow in stages, will be polished and proportioned well, and will be born as Adam with all the super functions of knowledge, reason, and language, it will be done accordingly.

For the creation of first life or first living cell, the required material was assembled and the process started at the opportune time. God had pointed out to the location of the place being the 'Ocean filled with fire' (52: 6) where the event was about to take place. Here the first brick of life was laid that culminated in the creation of man. German scientist Gunter Wachtershauser (1990) was nearer the truth when he hypothesised that life originated near sea floor in a hydrothermal vent.

The site should be in a shallow sea where clay transported from land may have been deposited by rivers. Such a site shall be fully equipped with all the items required for the formation and function of a cell. The following items necessary for the creation of life may have been assembled in one place in due course of time to achieve the goal:

1. Ocean water should have suitable salinity or the salts should be present in the same ratio as in the liquid contained in the human body.

2. All the necessary 18 elements found in human body are: silicon, aluminium, magnesium, iron, calcium, potassium, sodium, oxygen, hydrogen, carbon, chlorine, iodine, manganese, lead, copper, silver, and zinc.

3. Heat energy available from an active volcano.

4. Methane and ammonia, the two most important gases that help the formation of amino acids were present in the smoke released by the volcano.

5. The atmosphere of the earth during this period had no free oxygen. Under ocean water too, in the presence of heat and volcanic eruptions, the environment was totally devoid of any free oxygen.

6. Clay baked by volcanic heat was available as the most suitable medium for polymerization of amino acids. Scientists today recognize the importance of clay as it provides a catalytic surface for chemical reactions and the quick formation of complex compounds.

It is not known how the first living system came into being and what it was like. One can however speculate that the actual origin of life from the non-living chemical compounds took place in two stages. Firstly, large organic molecules like proteins that had synthesized abiotically changed into the first primitive living system called ebionts or protobionts.

Protection of Newly Formed Life

The whole process of the chemical origin of life was completed in about 800 million years. During that period the conversion of inorganic substances into protein and then the formation of prokaryotes had faced innumerable ups and downs. Adverse natural conditions had presented hindrances at every stage in the progress of cell formation. It was not only protein formation that was at stake, the newly formed living matter too was open to danger. One should however not forget that the situation was being handled by God, the Planner, the Executioner, and Stickler to Rules, the Perfectionist, and the Creator. The act was being done under the watchful eyes of this Supreme Being Who was creating it under the same rules of creation He Himself had framed. So instead of leaving the first living matter open to the vagaries of nature, He arranged for its protection and preservation too.

The first and most obvious way adopted was to put the newly formed cell into a protected place where it could be saved from the onslaught of local adverse environment. It could have been a secluded corner of the rift, shaft, or chasm in the ocean basin. It should be an area devoid of free oxygen that could have easily broken the fragile assemblage of protein. It should also be near a hot volcano so that the place remains heated and somewhat dry. The first single living cell formed during phylogeny may have been placed in a secluded, well-protected shaft near a source of heat in the ocean basin. Likewise, the living single cell, a nutfa (fertilized zygote) formed in ontology as a result of sexual reproduction is protected in a safe environment of the womb.

Besides keeping the first living cell in a safe place, it was also necessary to safeguard it from the onslaught of the harsh local environment. Molecules of the cell also needed seclusion and stability to perform their biochemical activity properly. The cell was also to be protected from any parasitic activity by any other cells to be formed in the future. All this was achieved by God through creating a cell membrane wall enclosing and protecting all the living cell molecules.

The other method of protecting the newly formed cell adopted by God was to incorporate a mechanism in the cell through which its mate of like nature was made (7: 189). The cell began to duplicate or true copies of it began to be produced quickly. Soon after the formation, the original first cell was divided into two daughter cells of identical characters. The process involves the cell wall growing inwards and dividing the plasma in the middle of the dividing cell. This wall separates the original cell into two equal daughter cells that possess characteristics identical to their parent cell.

The duplication of cells or the formation of the mate was performed quickly within 20-30 minutes. In order to safeguard the survival of the cell in the beginning, the duplication may have been very fast, probably taking only 20 minutes. In the process of division, the original cell grows to double in size and then splits into two identical cells or two daughter cells in 20 minutes. Thus in 20 minutes, one cell is made into two and in the next 20 minutes, these two cells become four. Further, in the other 20 minute,s the four cells increase to eight (Fig. 1). The total population of cells this way increases four times in merely an hour. The exponential growth makes the population increase into millions in a short span of time. If some millions are destroyed through environmental harshness, there still are left to further multiply and recover from the loss. This mechanism was enough to protect the newly formed, fragile life from extinction.

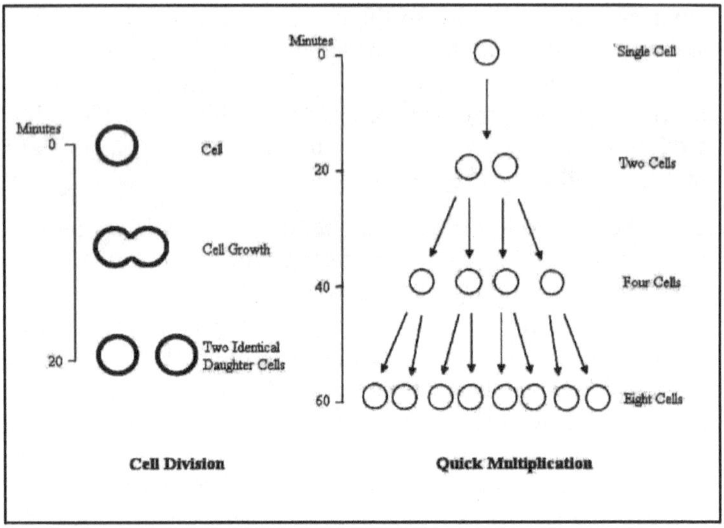

Fig. 1: Cell Division and Multiplication

Sustenance for new organisms

Safety of life was not the only problem faced by the newly formed single-celled organisms. In the beginning, they had been utilizing organic substances from outside for building their body and liberation energy. But with a further drop in earth's temperature the synthesis of organic molecules stopped. As a result, the supply of organic molecules formed earlier and accumulated in the 'sea broth' or 'sludge' was reduced through gradual consumption by chemoheterotrophs. But God could not let His creations die for want of food. He already had arranged the provision of sustenance for all the living organisms according to their requirement even before they were born (41: 10).

The provision of sustenance was arranged through the process of chemosynthesis of organic substances from inorganic raw material with the help of chemical energy. Such bacteria are still present in the form of sulphate-reducing bacteria, nitrifying bacteria or iron bacteria. The chemoautotrophs beside making their own food were also capable of making extra food for other living organisms.

The provision of sustenance for an increasing number of organisms was made possible by God Who introduced into the cell a green substance known as chlorophyll. This action of God was no less important, no less complicated, no less puzzling, and no less impossible than the creation of a living cell. The formation of green chlorophyll is the first brick on which the whole edifice of life in this world is erected. Southwick, C. H. (1972) is of the view that green plants combine carbon dioxide with water and, using the energy of sunlight and the enzyme systems of chlorophyll, can ultimately produce sugar and oxygen. It is the main source of energy for the proper functioning of ecosystems of the green plants. It lies at the base of providing organic food to millions of inhabitants of the earth. Chlorophyll in fact is the lifeline of existence and survival, but it also is the fact that no researcher has so far been able to find how green pigment suddenly appeared. It simply leads one to believe in the hand of God Who created it like His other creations including Man.

The enzyme system of chlorophyll uses the energy of sunlight and converts carbon dioxide and water into carbohydrates and other biochemical molecules. The conversion of light energy food or chemical energy by chlorophyll is called photosynthesis. The formation of chlorophyll and the process of photosynthesis started a mechanism through which regular manufacture of food energy became possible. Thus, food supply was made available for all times to come in one stroke by God by starting the process of photosynthesis. It had far reached consequences on the environment and the growing life. It formed the basis of sustenance for all life formations.

The food produced was used by the producers themselves to satisfy their own needs of respiration and metabolism and was also made available to satisfy the needs of other primary consumers. The first autotrophic bacteria that were able to form their own food through chlorophyll by the process of photosynthesis are called photoautotrophs. The earliest photoautotrophs formed about 3800 million years ago could only perform anoxygenic photosynthesis as they obtained the reducing power not from water but from the sources like and hydrogen (H_2), hydrogen sulphide (H_2S), thiosulphate (S_2 O_2--3) and some others. In this process, no oxygen was produced and they were therefore still anaerobic.

At a later stage about 3500 million years ago oxygenic photosynthesis evolved. Oxygen releasing photosynthetic organisms were cyanobacteria or blue-green algae-like organisms. They used water to get hydrogen and release oxygen. Slowly oxygen began to accumulate in the atmosphere, changing it gradually from a reducing one to an oxidizing atmosphere.

The source of sustenance however was not created to function in an erratic way. The amount of primary productivity follows the general rule promulgated by God. Food energy is not produced equally everywhere. It was to follow the pattern of demand and supply. It has luxuriant output in tropical moist forests where millions of species and individuals are present and need it, while it is smaller in polar areas where there are only a few users.

Another large-scale impact of photosynthesis was when the atmosphere and the earth readily agreed to God's command to develop mutual relationships (41: 11). As a result, the chemistry of the two altogether different spheres began to gradually change the primordial arrangement into a mutual interrelationship. It started with the condensation of minerals to form various layers of the crust and provide gravity to the earth. It was followed by copious rainfall over thousands (or millions) of years, and then the creation of life at an appropriate time.

The greatest advance toward the interrelationship was made by photosynthetic organisms that began to release oxygen into the atmosphere. In the beginning, the process was slow but gradually and steadily the oxygen began to accumulate in the atmosphere. Methane and ammonia started to disappear after being oxidized and carbon dioxide and nitrogen were formed.

Carbon dioxide further accelerated the process of photosynthesis as it provided the raw material for its function while nitrogen acted as a neutralizing force for excessive oxidisation. In due course of time the composition of the atmosphere changed from a reducing atmosphere to an oxidizing atmosphere. Free oxygen in the atmosphere combined with a number of elements present on the surface of the earth and converted the same into oxides.

The other elements essential for the survival of organisms like carbon, hydrogen, nitrogen, phosphorus, sulphur, etc. also began to develop co-relationship between living and non-living components of the earth and its atmosphere. Plants had grown as a result of the accumulation of photosynthetic products like carbohydrates and fat that made their leaves and other body structures. Plants were thus the first that were actively involved in bringing the earth and atmosphere closer to each other and exchanging chemical elements and mineral nutrients amongst them. The primordial surface of the earth was devoid of carbon and nitrogen, while phosphorus and sulphur were commonly present. The plants took CO_2 directly from the atmosphere by photosynthesis while nitrogen was taken through the roots. Phosphorus in the soil and sulphur derived both from soil and atmosphere were present for use. Most of the chemical elements or mineral nutrients are recycled within the earth and the atmosphere.

The earth shows its obedience to God's Command of functioning together with the atmosphere by releasing into the atmosphere some of the gases produced on it. The respiration of organisms or the decay of the organic matter releases carbon dioxide (CO_2), methane (CH_4), carbon monoxide (CO), and nitrogen oxides (NOx). Some gaseous compounds are also released from the earth. They are hydrogen sulphide produced in swamps and bogs, dimethyl sulphide from aquatic plants, and sulphur dioxide coming out from volcanoes. All the atmospheric nutrients are at some time or the other given back to earth through rain or snowfall or through dry deposits.

With the help of green pigment, God arranged for the sustenance of life for all times to come, formed the atmosphere and the earth to work in unison, and made conditions favourable for the survival of life through the transfer of mineral nutrients between their inorganic and organic components.

All these achievements of photosynthesis were side by side changing the chemistry of the chaotic atmosphere and arranging it into a functional unit.

The lower layer was made favourable for the survival of all the lifeforms by maintaining oxygen (20.90 %), nitrogen (79.07 %) and carbon dioxide (0.03%) in a fixed proportion. The ultraviolet light changed some of the oxygen into ozone that form a lifesaving protective layer in outer atmosphere. The presence of ozone layer is a boon to plants and animals and particularly to man, because it protects them from harmful effects of ultraviolet rays from the sun. This protective layer filters the incoming solar radiation by absorbing harmful ultraviolet rays and permitting only those light rays to reach the earth that are beneficial to the survival and continuation of life. Only visible light was then made available as the main source of energy. Apart from this safeguard, the protective layer helped in keeping the mean surface temperature of the earth at about 17°C. It would have been about –18°C had the outgoing infrared surface radiation not been trapped by CO_2, H_2O and ozone (O_3) (Cook Robert B., 1984). With cool visible light and favourable atmosphere, the lower heaven provided ideal conditions for the growth and survival of organisms (Cook, Robert B., *Environment: Science and Policy for Sustainable Development*, V01. 26 (7), 1984, pp. 10-40, cf. Southwick, C.H., *Global Ecology*, 1985, p. 70).

The supply of free oxygen in the atmosphere as a result of photosynthesis created inhospitable conditions for the survival of anaerobic organisms. Every increase in free oxygen started destroying the prokaryotes (anaerobic organisms). At a time when they were on the verge of extinction, God saved some by introducing a new method of aerobic respiration. Instead of killing, oxygen became the means of survival for the organisms. The mechanism of aerobic respiration not only saved organisms from extinction, but it was more efficient as the aerobic organisms released several times more energy than prokaryotes.

The change in atmospheric composition that had started with the formation of green pigment in the cell had its major side effects too. Henceforth, in the presence of oxygen, carbon dioxide, and nitrogen, there was no possibility of life originating again from inorganic substances. The engine of life was to continue in the words of Louis Pasteur (1864) *omne vivum ex vivo* (life comes from pre-existing life).

Suitable conditions for the origin of life, i.e., the presence of dead inorganic material in the form of clay and water together with a favourable atmosphere were made available 3900 million years ago. As the life started around 3000 years ago, it means that it took almost 800 to 900 million years to create first life or the first single cell. The protocells collected more organic material, formed cytoplasm or cellular machinery and became first living beings. All the living organisms created on Earth possess similar characteristics because they have originated from a common source or from the first single cell.

Points to ponder as to whether it was the result of Evolution or the handiwork of God:

1. Who first created a highly concentrated nano-particle and then exploded it with great force and with a loud sound?

2. Who first formed Universe and is continuously expanding it?

3. Who created galaxies and stars including our own sun?

4. Who created inorganic minerals and water necessary items required for the formation of life?

5. Who gave life to dead inorganic matter?

6. Who created the first living cell which according to intelligentsia was an impossible task?

7. Who provided a secure and protected environment for the survival of the newly formed living cells?

8. Who introduced in the first living cell a mechanism to divide quickly into identical twins?

9. Who first created chlorophyll?

10. Who started photosynthesis or a technique by which first organisms prepared food for their own use as well as for other organisms.

11. Who created circumstances to change a reducing atmosphere into an oxidizing atmosphere?

12. Who started the mechanism of aerobic respiration and thereby changed anaerobic organisms into aerobic organisms?

He is God Who gave rise to an atmosphere conducive to originating first living cell from clay and providing its protection and growth in future

CHAPTER II

Sexual Procreation

The practice of aerobic respiration helped the development of a refined respiratory system that, together with photosynthetic structure and other well-defined cellular parts, resulted in the growth of prokaryotes into eukaryotes. There are two views about the origin of eukaryotes. According to Raff and Mehler (1972), they were formed from prokaryotes through mutation and development of organelles. Lynn Margulis (1970) on the other hand thinks their mitochondria and chloroplasts originated as symbionts when various prokaryotic cells started living inside the cells.

A remarkable development was the formation of membrane-lined organelles and the creation of nucleus that produced eukaryotic cells. Creation of nucleus was a big jump in the development of new life forms. The event took place about 1700 million years ago that led to the formation of eukaryotes.

The formation of nucleus is an important achievement as it helps in introducing variety in newly formed organisms. Nucleus, in fact, is the double membrane-protected control system that contains DNA or hereditary information of the cell. In eukaryotes, the DNA is present in several pieces attached to a number of chromosomes.

The early eukaryotes continued to multiply asexually like the prokaryotes. During asexual multiplication, the parent cell is divided into two daughter cells. This is done through the process known as mitosis. Under this process, the DNA within each individual chromosome is divided into two identical chromatids. The divided chromosomes move to the centre and then the two chromatids from each chromosome are separately moved to the opposite end of the cell. The opposite ends possessing two sets of chromatids are separated into new cells. Each daughter cell thus produced asexually is similar to the parent cell and contains the same number of chromosomes and has identical DNA. Owing to identical DNA, no hereditary change is possible in the new cells formed asexually.

When unisex animals were created, they produced either male germ cells known as sperm or female germ cells referred to as ovum. Later on, a great change took place when God altered the process of procreation and henceforth organisms began to reproduce sexually through the process of interaction between male and female germ-fluids (76: 2). In this process of sexual reproduction known as meiosis or reduction division, homologous chromosomes come together, exchange some genetic material move to the opposite ends and then the cell divides into two parts. Each chromosome and the cell are further divided into two, thus making together four sex cells, each having half the normal number of chromosomes in a new combination.

When a male sperm infiltrates into a female ovum the two sex cells 'fuse' or fertilize to form a zygote with a normal number of chromosomes. The process of meiosis and fertilization are opposite functions but side by side are complementary to each other. Meiosis or division reduces the number of chromosomes in half, but fertilization doubles the number so that the zygote attains the same number of chromosomes as that of the parent cell.

Sexual reproduction through meiosis and fertilization has had far-reaching impact on the variety and growth of future organisms. At the time of meiosis or formation of four sex cells a number of re-arrangement of chromosomes takes place and genetic material 'cross over' between homologous chromosomes may occur. The fertilization of genetically re-arranged chromosomes through gametic meiosis introduces new traits in the newly formed zygote.

The process first originated in eucaryotes and started after the creation of organisms from clay a second major stage, recognised by God as *Nutfa* (zygote) Stage.[4] It enabled the eucaryotes to develop faster and with more variability. Hereditary change became possible and new life forms began to be created. Innumerable variations in the new eucaryotes produced this way are because of: i) Random arrangement of chromosomes during meiosis, ii) 'crossing over' or introduction of new gene linkages in chromosomes, and iii) random genetic union or 'fuse.'

By sexual reproduction, God started an automatic process of introducing innumerable variations in the coming generations of organisms. It also provided a stimulus for accelerated growth and development of new types of not only eucaryotes but all the organisms formed sexually.

The unicellular organisms created as a result of gametic meiosis are very small and microscopic and are complete functional units. The individual cells have their own independent identity and carry out all the life activities within themselves. They live alone and never interact or merge with each other. They may however come close to others and make colonies.

Unicellular eucaryotes created through *nutfa* are considered by God as the great leap in the development of life. He regards sexual reproduction as the second major stage recognized as

---

[4] Sexual reproduction is mentioned in many Verses of the Qur'an alternatively as a conception of embryo in the womb or as a means of creation in nature at large. The distinction between the two types of narration is evident from the fact that if the Verse mentions creation from dust and then from sexual mechanism, it points to general stages of creation. In case simple sexual formation is referred to, it may stand for fertilization in the womb.

nutfa just after the first major breakthrough of creating life from clay. All the major unicellular organisms created from *nutfa* have ruled the earth as the only organisms for about 200 million years before other forms were created. Such unicellular organisms are known as protists by scientists.

The structure of unicells is a double envelope system. The protoplasm is surrounded by a plasma membrane and the internal organelles too have distinct membrane covering. The hereditary or genetic material is high and is present as organized linear DNA.

These organisms of various types either manufacture their own food or obtain it directly from their immediate environment. Autotrophs make their own foods through the process of photosynthesis while others are phagotrophic in nature and ingest nutrition from their nearby surroundings. Others may be saprotrophic or parasitic and feed through absorption from the organic remains of others. A few others may be mixotrophically obtaining their nutrition through two or more processes mentioned above.

They are aquatic organisms and may live in back waters, float on the surface, or may be present as benthic biota deep inside on the ocean floor. If they remain on the surface of water their respiration is aerobic, but at the bottom, it is anaerobic.

Only a few of the unicellular organisms are sedentary and just float in water, but most possess the power of locomotion and move from one place to the other. The movement is achieved by diverse mechanisms. Some creep over the floor with the help of pseudopodia or false foot. There are protrusions of protoplasm that are thrown out as false feet in the direction of movement and are withdrawn from the rear end. This function makes the organisms creep on the solid sub-stratum. Others have been provided with hair-like protoplasmic overgrowth of flagella and cilia and propel themselves through undulatory movement to directions opposite to that of the strike. The movement by flagella or cilia is most common in a number of unicellular organisms. Wriggling movement and mucilage propulsion are some other means of locomotion seen in these organisms.

Protists are recognised as two subgroups on the basis of the presence or absence of green pigment or chlorophyll. Those with chlorophyll are called algae and belong to the plant kingdom while those without chlorophyll are protozoans of the animal kingdom. Animals are the main focus of description by God to show how man was created through different pathways of creation. Therefore, only animal protozoa out of various *nutfa* organisms are described here briefly.

Protozoan protists are microscopic unicellular organisms occurring in aquatic habitats whether marine, salt water, or fresh water. They live as unicells alone or may form colonies. In colonies too they maintain their individuality and separate identity. Protozoa are regarded as animals because of their animal-like nutrition and locomotion. The locomotion is caused by finger-like pseudopodia, whip-like flagella, or short and hair-like cilia. The cell wall is absent and therefore the body is naked. The body commonly contains a single nucleus but a number of them also have a wall. The common mode of nutrition is holozoic (phagotrophic). In some form, it is saprobic, and has more than one nucleus bounded by a delicate membrane or a firm pellica.

There are many types of protozoan protists that differ from each other on the basis of their locomotion. Some live freely in 'aquatic' habitats, while others live as commensals, symbionts, and parasites in other organisms. They move in water by moving their specially formed flagella. The number of flagella may range from one to eight or nine in different species.

Fossil records

Body parts of organisms are preserved in fossilized form by natural processes. They are commonly formed in sedimentary rocks that are formed of material deposited in oceans. Any organism buried with the sediments is easily fossilised. *Nutfa* organisms lived in water and therefore had an ideal location for getting buried and converted into fossils. However, since unicellular organisms is soft-bodied, they could not be preserved as soft parts decay and disappear easily and quickly. Such organisms therefore have left no fossil records. Some traces or impressions of unicellular organisms have, however, been noticed in a few places and that gives an idea of their form and when they were created. Southwood (2003) in his book '*The Story of Life*' states that the oldest fossils of Agrypnia protists were found in 1992 in rocks formed 2100 million years ago. These and other filamentous fossils are present in more recent rocks (around 1200 million years old). As for younger and recent fossils, Southwood writes: "From about 1700 Mya numerous, generally spherical, fossils occur. The earliest ones are simple but those from younger deposits often have elaborate cell walls—sculptured or with spines. They are usually interpreted as the cysts of primitive algae. However, unlike some the bacterial microfossils, they are not easily placed in modern groups... The oldest examples are small, a fraction of a millimeter. One, *Chuaria* , can be as much as a centimeter across and is the largest organism found in the fossil record up to this date... Although acritarches, especially the large forms, become rarer in the fossil record after about 900 Mya, recent discoveries in India and China reveal that q occurred until 570 Mya. A fossil which may be that of red seaweed (bangiom orphan alga) has been found in the Hunting Formation, Somerset II, Canada (750-1250 Mya). This apparent diversity of protists during the period 1100-900 Mya probably reflects the evolution of meiosis, that is, of sex." (Southwood, T.R.E., *The Story of Life*, 2003, pp. 20-30).

Points to ponder as to whether it was Evolution of the handiwork of God:

1. Who created the first nucleus in the cell?
2. Who protected the nucleus by a double membrane envelope system?
3. Who started sexual reproduction through meiosis and fertilization?
4. Who first formed gametes or Nutfa?

5. Who first created a mechanism for introducing variability in hereditary traits in future generations?

No one except Almighty God has the power to give life to clay and introduce genetic variations in developing body parts.

CHAPTER III

Multicellular Clinging Entity

This stage of creation in the Journey of Life holds an important position in the eyes of God as the first five Verses revealed to Prophet Muhammad include the information that man was formed from *alaq* (96: 1-2). Being part of the first revelation, this information carries the weight of what is being told. It is to highlight that man, in the beginning, was a lowly creature like an alaq and then God perfected him as a full-fledged intelligent being. The importance of this stage is further strengthened by God when He regards it as a part of one of the seven stages of the journey of life (22: 5).

The Arabic word alaq literally means a leech. It is unique in the sense that the word represents all the characteristics of a leech in one, i.e., in shape it looks like a leech, is a blood clot full of blood, and like a leech is a clinging entity that attaches or is suspended to a body part. Human embryo a week after fifteen days of conception appears like an alaq, a leech.

Alaq is a formative phase when the hand of God is seen in the creation of various biological traits in the organisms. During 150 million years of this stage many innovations were introduced in the structure and function of the creatures.

This is the stage when for the first-time leech-like multicellular creatures with soft body or without internal skeleton were created. Zoologists in their animal classification regard it as the beginning of the stage of invertebrates or the stage when animals were born without skeleton.

It was also for the first time that red blood, the indispensable lifeline for man, was created in the organisms that belonged to this most important stage in the life history of creation. Man has passed through this stage in his journey towards the goal to be known as a final product of creation: "...then afterwards We formed out of it (alaq) a final creature (man), so blessed be Allah, the best to create." (23: 14).

Multicellular Organisms

The unicellular organisms that existed in Nutfa Stage were maintaining their individual identity whether living alone or in groups. Other eukaryotic cells when living in groups began to interact with each other and became interdependent. This major change, though crude, heralded the onset of Alaqa Stage when organisms functioning alone began to work interdependently and this function made it a crudely formed multicellular group.

The cells were of different types but they functioned in unison. This early stage of closely-knit cells of various types is still seen in certain sponges. The sponges have a cell aggregate body plan but tissues do not differentiate. They are formed of two germinal layers but they are not the real cellular layers of the body wall of the animals. These are two imperfect cellular layers enclosing a non-cellular gelatinous layer in between. Several types of cells occur in the wall of sponges. They are pinacocytes and phorocytes that make the outer body wall called pinacoderm, and secondly choanocytes or collar cells that make the inner cellular layer. The choanocytes bear a contractile collar and a single long flagellum. They create water currents by the movement of their flagella. Although sponges show a cellular level body plan with no tissue differentiation, their work in unison may be taken as a rough beginning of multicellular colonies.

Two and Three Cellular Layers

The body wall that started with the arrangement of several types of cells into germinal imperfect layers in sponges was converted into cellular layers. The early simple multicellular animals were thus created with two perfectly created cellular layers of the body wall. These diploblastic animals, for example, coelenterates had an outer epidermis and an inner gastrodermis and together enclosed a gastrovascular cavity. Later on, the animals with three layers or triploblastic animals like flatworms or acoelomates were created. These animals had outer ectoderm, inner endoderm, and in between a mesoderm. All three layers enclosed a digestive tract (Fig. 2).

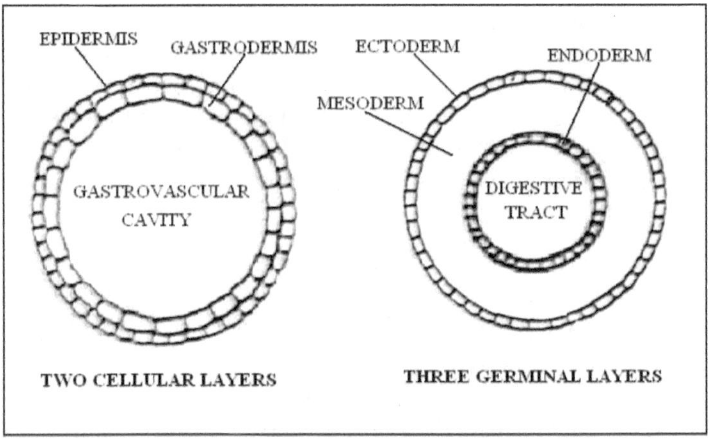

Fig. 2: Cellular and Germinal Layers

The creation of three layers was the hallmark of Alaqa Stage and all the animals henceforth were formed with three layers. These are called germinal layers as all parts of the body of adult animals develop from these layers. The cells of each germ layer produce a specific set of organs in all the triploblastic animals. The structural organization of the body from a cellular level like in sponges through tissue level arrangement of specialized coelom (body cavity) cells of coelomates to the final organ system level of the animals was a long jump in the improvement achieved during Alaqa Stage.

Another major breakthrough in the creative activity of God during the Alaqa Stage was the creation of a body cavity known as a coelom within the middle layer. The coelom separates the digestive tract from the body wall. It is lined by parietal peritoneum on the outer side and visceral organs are present inside the coelom. Perfect coelom was formed by splitting mesoderm into two parts (Fig. 3).

The presence of coelom is an essential improvement for the proper function of various organs. The digestive and circulatory system organs can function properly in the space provided by coelom. The coelom fluid itself acts as a simple circulatory system transporting food, water, gases, and wastes around the body. The fluid provides a moist environment to visceral organs and protects them from any friction or outer shock. Lying within the coelom, the gut can grow independently without any space restriction. It can even grow much longer than the main body by coiling and folding within the coelom.

The formation of the body cavity was a gradual process. Animals that were first created with three layers of the body wall were without any body cavity. The only cavity present in the body was the digestive tract. Examples of three-layered animals without coelom are platyhelminths' or flatworms.

Roundworms or nematodes were the first to be created with a partial formation of the coelom. They are called pseudocoelomates wherein the so-called body cavity or false body cavity is derived from blastocoels and is not lined by the peritoneum. Mesoderm is present like pouches in between ectoderm and endoderm. The formation of coelom was later on perfected and annelids, also known as coelomates were created with a true body cavity. The coelom known as schizocoelom is lined by mesothelium with an outer parietal and an inner visceral layer. The fluid present in the coelom is known as coelom fluid.

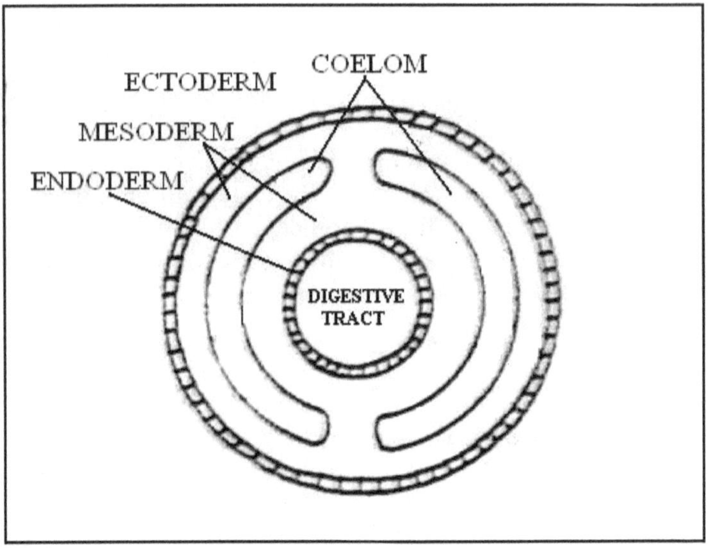

Fig. 3: Body Cavity or Coelom

Digestive Tract

Before the formation of coelom, animals had passed through a cell aggregate plan to a blind sac plan. The first animals like sponges had meagre cell differentiation and the body were just an aggregate of cells that were not organized even into tissues. The condition was later improved and a sac-like single cavity was created inside the body. The coelomates and flatworms created with such a blind sac plan had to use a single mouth for both taking the food and also for throwing out the fecal matter.

The creation of the coelom changed the arrangement drastically. The body now had two tubes: an outer body wall separated by coelom from the inner digestive tract. The formation of coelom at the embryonic stage started from the blastopore from the front where the mouth was formed for ingestion of food and the anus was formed at the end for removal of excreta. This arrangement is seen in roundworms, annelids, mollusks and arthropods of the Alaqa Stage.

Red Blood

The transport of material within the body of animals is made possible through a fluid tissue known as blood. In early animals, the blood just flowed in open spaces, the sinuses, and directly bathed the living cells. Later on, in a few cases of arthropods and in most cases of mollusks, a semi-developed circulatory system was created where a heart and a few blood vessels were made. The heart forces blood by muscular contraction through the vessels into open spaces. The system was further refined in annelids when closed blood vessels with definite walls were provided for blood to move in a totally closed circulatory system.

Colour of the blood depends on the dominant metal element present in the body. It was copper containing blue coloured, hemocyanin in arthropods. Later on, iron protein was introduced in some annelids that for the first time provided reddish pigment haemoglobin in the blood. The red blood dissolved in plasma in the annelids, was perfected as true blood or haemoglobin inside RBCs in the mammals created at a future stage of development of life forms.

The creation of red blood and the creation of leech-like animals were important events and God emphasized the point by selecting them as representative characteristics that gave the name Alaqa to the Third Stage of life formations.

Soft Body

All the worms, Platyhelminthes, aschelminthes and annelids are soft-bodied alaq animals. The soft body is kept in shape because of high amount of fluid in the body is incompressible, resists pressure, and keeps the body intact.

A soft body however has its own disadvantages. It can easily be damaged, devoured, or put to death. Its movements, because of little foothold strength, are restricted and the animal cannot run away fast enough to protect itself from predators. God has however provided some of the soft-bodied animals with a protective body shield in the form of a hard external skeleton, like chitinous plates in arthropods or calcareous shells in mollusks. Some others like echinoderms were provided with thick, hard, calcareous spines to protect as well as threaten the predators. While some were given protection, some others were armed with special tools to damage or penetrate even this hard shield, e.g., claws and some biting and grinding hard parts in the mouth of crabs, and drilling or rasping parts in the oral cavity of snails.

Classification

All the soft-bodied animals are devoid of hard endoskeleton. But some of them possess a soft skeletons inside their body as body support. In view of the absence of a hard endoskeleton, all the animals of the Alaqa stage are known as non-vertebrates (invertebrates) in the scientific world.

The non-vertebrate Alaqa Stage has a dominant position in the pathways of life formation because during this period new beginning was made for the functional formations of the complex organ system. God introduced gradual improvements in all the successively created soft-bodied alaq species. The improvements incorporated in eight major phyla of non-vertebrates arranged by Zoologists are as follows:

1. Porifera, rudimentary animals with simple structures, also known as sponges, are the first to be created in the Alaqa Stage. They are formed of many cells but all function independently with very little cooperation with each other. The cellular organization is there but they have no tissues or organs. The Porifera are considered as a link between plants and animals, but they have been created with some improvements from the earlier protists. They have a larger multicellular body with a greater division of labour among the cells. The reproduction is through female and male gametes in the form of spermatozoa and the zygote develops by segmentation instead of mere growth.

2. Cnidaria or Coelenterates are named owing to the presence of special stinging cells, endoblasts that are used for both defensive and offensive purposes. They live alone (singly) or

form colonies. Their body forms may be polyps or medusa. Polyps are cylindrical, usually fixed, may be solitary or colonial, are cup-shaped, and look upwards. Medusae are free swimming, solitary, umbrella-like creatures facing downwards.

Coelenterates are privileged amongst animals to have been created for the first time with a rudimentary nervous system and with sensory cells. As a result, they work in better coordination than their predecessors. On the whole, cnidarians have certainly been created with some other improvements over sponges in the form of theirtissue-levell organization, presence of digestive cavity with intercellular digestion, and formation of the ductless male organ.

Examples of Phylum Cnidaria are hydra, jellyfish, and sea anemones, and corals.

3. Platyhelmithes or Flat Worms are flattened animals possessing a soft body of bilateral symmetry. They hold a unique place in the line of creation as God for the first time introduced into them a third germ layer that became an important feature of all the animals. With the creation of the third layer, mesoderm, between ectoderm and endoderm, they became the first triploblastic animals. The cells are organized into a primitive organ system packed in tissues between the body wall and digestive tract. The body is enclosed by a single-layered epidermis and presents a sac-like appearance.

Platyhelminthes are the first animals to be provided with a definite head that also contained sense organs. The brain and two longitudinal nerve cords provide a ladder-like nervous system. Gonads are armed with gonoducts that were absent in earlier creatures. A male copulation organ has also developed for the first time in these animals but fertilization is internal. With many firsts, flatworms have certainly been created as advanced animals than cnidarians.

4. Nematoda or Nemathelminthes or Round Worms are cylindrical with tapering ends on both sides. They are triploblastic with three derm layers and have an organ system

level of organization. The first complete digestive system was created in these animals. There is a mouth in the anterior region for intake of food, and an anus or cloacal aperture in the posterior for throwing out the faeces. The intestine is present to allow proper movement of food without mixing it with faeces that is removed quickly from the anus. Their posterior part of the body extends beyond the anus as a tail, which is a unique feature in invertebrates.

Coelom in nematodes is present between the body wall and the digestive tract. It is however not lined with a clear wall of demarcation and appears only as a false cavity known as pseudocoelom. It may be the beginning or the initial stage of the formation of the coelom, but it is a great step in the journey of creation.

Nematodes are the first to be created with separate male and female sexes which perform bi-sexual reproduction. Males are smaller than females and possess pineal specules for copulation.

Nematodes were created with improvements over platyhelminthes, and all the new formations were of great consequence for future creations. These improvements of far-reaching consequence introduced for the first time in roundworms as mentioned earlier are the formation of a complete digestive system, beginning of a fluid filled body cavity, and separate male and female sexes.

5. Annelida or Segmented Worms are soft-bodied, cylindrical, and triploblastic worms of bilateral symmetry. Their body is divided into segments or metameres that are visible externally in the form of ring-like grooves or annuli. The segments are separated inside their body by a partition wall known as septa. The body cavity that was false in the nematodes is now created as true coelom in the annelids. It is schizocoel or formed by splitting of mesoderm and is divided into compartments filled with coelomic fluid. The digestive tract is straight, extends through the entire body, and carries out extracellular digestion. A distinct head is present in the

anterior end of the body and possesses feeding, sensory, and nerve centres.

An important development in the annelids is the creation of a closed circulatory system. The heart pumps blood through dorsal and ventral blood vessels and their branches to every part of the body; skin, muscles, intestine, and other organs. Red blood has been created for the first time in annelids from haemoglobin and erythrocruorin. It is present dissolved in the plasma as RBCs are absent. While all the other annelids have blood vessels, leeches are an exception wherein the blood vessels are absent. The red blood in them is filled in coelom where it is known as haemocoel.

On the whole, Annelida were created with certain improvements over round worms. These advancements are - Formation of metameric segmentation; occurrence of a complete digestive tract; presence of a definite head; creation of a true coelom; existence of a closed blood vascular system; and formation of haemoglobin.

6. Arthropoda or the Animals with Jointed Legs and Appendages have bilateral symmetry, possess three germinal layers and thus are triploblastic, and contain organ system level of organization. Their body is covered with an exoskeleton of chitinous material which is externally segmented but internally segments are not separated. The segments are grouped into two regions; head with trunk and abdomen or into three regions; head, thorax and abdomen. A distinct head is present which has mouth parts and various sense organThe mouthuth has moving appendages that help in capturing, chewing, sucking, etc. Eyes are generally compound possessing several lenses. Each lens captures image and a number of images thus formed to help these animals notice even the slightest movement in the neighborhood. Arthropods have a reduced coelom present in a few pockets near the reproductive organs. The digestive tract is complete, digestion is intercellular, and the alimentary canal has three parts: a foregut, mid-gut, and a hind-gut. Some or all segments have jointed appendages that have been variously modified to

suit their specific use in clinging, swimming, walking, jumping, feeding, etc.

The arthropods have been provided with several improvements over annelids. They have paired jointed appendages for performing a variety of functions and a tough jointed exoskeleton for protection and muscle attachment. A distinct head is present in all the species. They also possess highly developed sense organs such as statocysts, auditory functions, taste receptors, and more especially, compound eyes.

7. Mollusca or Shelled Animals are soft-bodied triploblastic animals' bilateral symmetry usually encased in a hard shell; slugs and squids lack an external shell. The body has three parts: a head containing mouth, eyes, and tentacles, a ventral structure, foot used for swimming, burrowing, and creeping, and a dorsal visceral mass. The underbelly is formed of a muscular, vascular, and glandular fold called mantle, which though thin, is hard enough to protect soft organs inside. The mantle secretes a limy substance that forms a hard covering or shell enclosing the soft body.

Coelom in mollusks is greatly reduced. It is present in small cavities near the heart, kidneys, and gonads. The mouth is provided with rasping tools called a radula that contain transverse rows of teeth. A three-chambered heart contains 1-2 auricles and 1 ventricle. Blood is generally blue in colour containing copper and is known as haemocoel. Gills are provided for respiration.

The improvements in mollusks over annelids are in their distinct head formation, presence of a three-chambered heart, and well-developed gills.

8. Echinodermata or the Spiny Skinned Animals are triploblastic having an unsegmented body of radial symmetry. The body is covered with thick calcareous spiny structures to protect them from any attack by other animals. They show the beginning of an endoskeleton in the form of small calcareous

plates originating in the mesoderm and embedded in the dermis.

A true enterocoelom is present in echinoderms. The fluid-filled in the cavity is used for the transport of food and gas. A part of the coelom is converted into unique water-filled canal system. A perforated plate permits the entry of water into the system that goes through finer canals into various body parts and helps the animal in locomotion and respiration.

Echinoderms are created with such features that make them stand at the turning point from one stage to the other. They are formed with a simple endoskeleton that brings them closer to higher-level vertebrates. They are deuterostomic in nature, i.e., their mouth is formed opposite to the blastopore which forms the anus. This is the quality of chordates too. Echinoderms, therefore, are the creatures that stand at the dividing line between invertebrates and vertebrates.

Age

Alaq animals are soft-bodied and do not possess hard, resistant endoskeleton. The soft parts quickly decay, are easily destroyed, or are eaten up by other predators amongst them. Such animals, therefore, are rarely fossilized. Moreover, these animals are marine in origin and reach the bottom of the sea soon after their death where they are buried under sedimentary deposits. Here they are liable to be destroyed easily due to the pressure of overlying rocks and greater heat inside. The possibilities of fossil formation of alaq animals are thus very rare.

A chance discovery of older animal fossils that resemble jellyfish of alaq group of animals was made in Ediacaran Hills in South Australia in 1946 by Reginald Claude Sprigg. Later collections from this site confirmed that these fossils also existed in the later period of Pre-Cambrian era. These animals were widely distributed over the world as their fossils have reportedly been found in Namibia, China, Russia, northern Europe and Newfoundland (Southwood, R., *The Story of Life*, 2003, p. 39).

On a close scrutiny of all the fossils present worldwide of Ediacara Age (590- 545 Mya), Southwood recognized three types of fossilized alaq fauna. First was a large group of circular animals like jelly fish and other related animals of radial symmetry. The other group consisted of fond-like forms that resemble some modern sea-pens. The third group was of most varied animals collectively known as worms. They were triploblasts having three germ layers and their body had bilateral symmetry.

The closing period of Alaqa Stage witnessed a unique creation. The soft body was provided with a protective cover of hard and tough material, a calcareous shell. The exoskeleton present in the form of claws and mouth parts or as body cover not only helped in killing organisms and protecting individuals but it, being hard, could also easily be fossilized. As a result, fossil remains of the closing Alaqa stage animals are available abundantly in Cambrian Period (545-505 Mya). These fossils indicate existence of a rich fauna with a wide variety of body structures during this period which is therefore known as 'Cambrian explosion'. A variety of animals with new arrangement of body structure continued to be created increasingly until Ordovician Period (505-438 Mya). The creation of a large number of animals with new body plans slowed down after the Cambrian, instead the number of families of the then existing organisms began to increase (Southwood, T.R.E., loc.cit, p.47).

The fossil records available so far reveal that the Alaqa stage (600-450 Mya) existed for about 150 million years.

Points to ponder as to whether it was Evolution or the handiwork of God:

1. Who formed first crudely formed multicellular organism?
2. Who created cellular layers?
3. Who made for the first time three layers of the body wall or three germinal layers?
4. Who created for the first time a body cavity or coelom?

5. Who made the sac like body cavity, a digestive tract with single opening used for taking

food and also for throwing out faecal matter.

6. Who improved the digestive tract with two openings, one a mouth for ingesting food and the other an anus for disposing off faeces.

7. Who first created a semi-developed circulatory system with a pumping apparatus, the heart and a few blood vessels?

8. Who created vessels for a closed blood circulatory system?

9. Who first created red blood or haemoglobin dissolved in plasma?

10. Who first created a body shield or hard exoskeleton or calcareous shell or spines?

11. Who first created a tough exoskeleton?

12. Who first created claws, grinding hard part in the mouth or drilling and rasping parts in the mouth?

13. Who created ductless gonads for the first time?

14. Who first created a definite head with brain, sense organs, and nerve cords?

15. Who first created a 3-chambered heart?

16. Who first created gonoducts with copulatory organs?

17. Who first created separate male and female organs?

The answer can only be God.

CHAPTER IV

Appearance of Chordates

Mudgha (a lump of flesh) in Arabic literally means a 'chewed substance'. Any piece of chewing gum or raw or half-baked meat on chewing gets on it the impressions of teeth mark as a curved line. The teeth impressions are indicative of a curved line of disjointed vertebrae. The formation of a curved rod-like item that is partly formed and partly unformed is the characteristic feature of animals created during Mudgha Stage (22: 5). It is the great stage when God started creation of skeleton bearing animals. Zoologists call it the stage of chordates when animals bearing a skeleton rod known as notochord started to appear on the scene around 540 million years ago.

The notochord is disjointed elastic but stiff rod created as a mid-dorsal line between the alimentary canal and the dorsal hollow nerve. But in the animals created at the close of this stage it peters out in the adults. It begins to be gradually replaced later on by partly formed and partly unformed vertebrae. It finally disappears from adult animals at the end of Mudgha Stage when it is totally replaced by a vertebral column or backbone comprising several vertebrae.

Notochord provides support for muscle attachment and works as anchor for body movement. It is the early formation of a fixed real skeleton. The endoskeleton had started to form in alaq animals like echinoderms but then it was in the form of moving and non-living calcareous plates. In mudgha animals the endoskeleton was made to grow and was a living part of the body. The endoskeleton in the beginning was limited to a few vertebrae of soft cartilages but with the passage of time by the end of Mudgha Stage it became a completely cartilaginous vertebral column. At the close of this stage bony skeleton had started to be formed and animals with 'partly formed and partly unformed' or a mixture of cartilaginous and bony structure were created.

An important feature of mudgha animals is the creation of gill-slits. These are a series of apertures, or clefts formed from the wall of pharynx near the mouth to the exterior of the body. The gill-slits are used for respiration and feeding. Like a kind of endoskeleton created at the close of Alaqa Stage, gill slits too were created in arrow worms and some related echinoderms, but at the onset of Mudgha Stage, such animals were formed that possessed beside gill slits a structure somewhat similar to notochord and thus were different from alaqa animals. Appearance of animals with gill slits heralded the beginning of Mudgha Stage. The gill slits are characteristic features of only mudgha animals and remained functional in the adults throughout their life. By the end of this stage gills began to disappear like notochord and such animals were created that possessed them only in the embryonic state but were devoid of gills later.

These animals are provided with a functionally better nerve cord. In the alaq animals the nerve cord was solid and ganglionated and was present in the mid-vertebral line in the coelom beneath the digestive tract. But in the mudgha animals it was made independent of coelom and was shifted to mid-dorsal position in the body and placed above the notochord (Fig. 4). It was now a single, tubular, fluid filled, non-ganglionated chord. As a result, more nerve cells could be accommodated and a more efficient and well-integrated nerve system was developed.

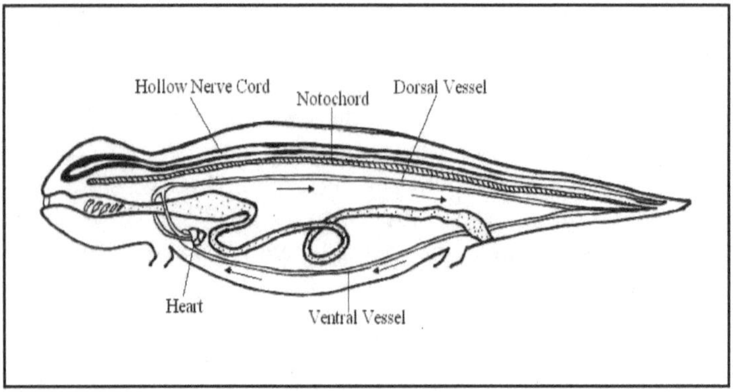

Fig. 4: Location of nerve and notochord in Chordates

The blood circulatory system in mudgha animals had become wholly closed and more efficient. The efficiency increased as a result of heart location in a ventral position as compared to the dorsal location in alaq animals. The circulation of blood also got reversed from its forward movement in the dorsal vessel and backward movement in the ventral vessel in the alaq animals to backward movement in the dorsal vessel and in forward movement in the ventral vessel in the mudgha formations.

One more unique feature of mudgha animals was the formation of fins as appendages. Earlier animals like arthropods too had appendages but they were in the form of segmented pores and were present in pairs. Fins are different in formation and typical to 'partly formed and partly unformed' character of mudgha animals, and are paired or unpaired. In the most primitive stage though they were absent but later appeared in the form of fin rays. Later animals were created with three unpaired fins like a dorsal fin, a ventral fin, and a caudal fin. At the close of Mudgha Stage some such animals were formed that possessed both paired and unpaired fins.

All the animals created during Mudgha Stage can be grouped on the basis of variations in their nature and structure into those belonging to either lower, middle, or upper Mudgha Stage.

Lower Mudgha Stage

Lower mudgha animals are recognized as Hemichordata, Urochordata, and Cephalochordata.

Hemichordata are most primitive animals that are more like a worm and are not formed with a thin notochord. Instead, a small hollow cord like outgrowth called stomochord is formed from the gut. The animals therefore are also called stomochordates. The presence of gill slits however brings them somewhat close to mudgha animals.

Hemichordates have bilateral symmetry, are born with three germinal layers, and have a true enterocoel. Appendages like tentacles and arms are present but they are not used for locomotion. A small dorsal tubular nerve cord is present in the collar region. Heart is not located in the ventral region like other mudgha animals and the circulatory system too is open. Tongue and acorn worm is an example of hemichordates that live in U-shaped burrows in shallow coastal regions.

Urochordata or Tunicata have sac-like body of varying shape enclosed by a thick tunic. Notochord is present in the tail of the larva or tadpole, hence it is known as urochordata (Greek: oure= tail and chorde= chord). Notochord however disappears at the adult stage. Likewise, a dorsal tubular nerve cord is present in the tadpole but it is converted to only a dorsal ganglion in the adults. The gill slits are present and are used for food intake and for respiration.

Cephalochordata have a fully developed segmented notochord extending from snout to the tail. Its existence has helped in giving them their name as cephalochordate (Greek: kephale= head, chorda= chord), although the real head is not formed. A complete notochord continues to exist all through the adult life. A dorsal, hollow non-ganglionated nerve chord is present from anterior part to tail but functions without brain which is absent. The nerve chord also persists throughout adult life. Enterocoel is present as a limited cavity just only to accommodate pharynx. The pharynx is large and divided into gill slits that open into the atrium and not outside. The gill slits are used for filter feeding only and not for respiration as the blood has no respiratory pigment. The fins are not paired. There is a long dorsal fin that extends for the whole length of the body. The ventral fin is short and is limited to posterior end. A caudal fin is present on the tail.

Middle Mudgha Stage

New features were introduced in the animals created during the middle Mudgha period. Most notable advances over earlier animals are the formation of cartilaginous endoskeleton, cranium, and a functional heart. The animals created at this stage are identified into four classes on the basis of their gradual improvement in certain characters. These classes are: Cyclostomata, Chondrichthyes, and now extinct Ostracoderm and Placoderm.

Cyclostomes are the earlier formation of Murgha Stage. With them a beginning was made in the creation of endoskeleton. The endoskeleton in cyclostomatous is however in the form of cartilage. Notochord is still present and persists all through its life. A beginning of vertebral column is seen in the animals but it is in the form of a small imperfectly formed neural arch over the notochord.

Mouth is circular and sucks the food. Teeth are formed like small horns on the tongue. Stomach is not formed and the intestine is folded. There are 5 to 16 pairs of gills placed in a pouch that is used for respiration.

Heart has been formed like S-shaped tube enclosed by a cartilaginous envelope. It has two chambers, one auricle and one ventricle. Red blood corpuscles are present, but they are irregular and nucleated.

Olfactory sac is single and median and opens into the outside with a single pore. There are also lateral line sensors to feel waves and currents in the water. The ear is membranous, each having 1 to 2 semi circular ducts. Fins possess a median fin supported by cartilaginous fin rays, a dorsal fin and a caudal fin. The reproductive system contains a single gonad without any conduct.

Notochord in Chondrychthyes is present in the embryonic stage but becomes a vertebral column at the mature stage. The body is formed of a jointed head, trunk and tail. The neck is absent and the head is directly connected with the trunk. Head consists of a proper cranium with ten pairs of cranial nerves. The olfactory sacs are vertical and large. The membranous ear lies inside and has three semi-circular canals. The lateral line sense organs are often well developed. The mouth is ventral to body axis and is split into a pair (lower and upper) of moveable jaws. The intestine is short and contains an anti-clockwise spiral valve or scroll valve that increases its surface area.

The respiration is carried out by 5-7 pairs of plate-like gills. Gill slits are generally without gill covers. The blood present in gills takes up oxygen from water passing through it or they breathe through gills. The respiration thus is possible only in the presence of water and the fish therefore cannot survive out of water.

The heart consists of two chambers and is ventrally located. The red blood corpuscles are oval, biconvex and nucleated.

There are two dorsal fins, one posterior pelvic fin and an asymmetrical caudal fin. The dorsal fins are generally paired and symmetrical.

The reproductive organs are visible outside. The gonads are paired and are connected with gonoducts.

The chondricthyes have a developed endoskeleton with cartilaginous vertebral column and cranium, paired olfactory organs, three semicircular canals in each internal ear, presence of a moveable pair of jaws, a pair of dorsal fins, and paired gonads.

Upper Mudgha Stage

The closing period of Mudgha Stage is the time when new innovations were incorporated in the animals that laid the foundation for future animals to be created in the next stage of life formations. Some of the cartilaginous vertebrae were made bony. Beside a partly or wholly bony structure of ventral column, the endoskeleton is also present in the form of a cranium that contains a better developed brain. The notochord is still present but in a greatly reduced form.

These animals have an improved respiratory system. Four pairs of filamentary gills provide a larger surface area for oxygen exchange than the laminar gills of cartilaginous fish. Gill slits are protected by gill covers.

Another innovation made in these animals is the presence of an extra apparatus for internal breathing. A gas filled swim bladder is formed that in some bony fishes is used as a lung for breathing. The bladder was mainly formed as a float or as a buoyancy regulator that enables fish to stay at a particular depth without spending energy in swimming. The swim bladder is filled with gases that in some fishes also help in breathing. Armed with this facility each fish is able to come out of water, crawl on the land surface for some distance, and breathe in the open through swim bladder.

The fins in osteichthyes are supported by cartilaginous or bony fin rays. There is one dorsal fin, but caudal fin is paired and symmetrical. The pectoral and pelvic fins are placed in the anterior position and help the fish to balance and provide breaks during swimming. A bony fish, Latimeria chalumnae or the coelacanth is created with all the fins except anterior dorsal fin that is formed as lobes. This fish also has breathing sacs and is able to move about on land easily with the help of lobed fins. Such lunged and lobed fin fishes are presumed to have been used by God to incorporate further changes and create animals that could live and breathe on the land surface.

The digestive organs include an anterior located mouth and the alimentary canal that opens out through anus. Cloaca is not present and the urinogenital aperture is separately located from anal aperture. The intestine is long, has sufficient surface area for absorption and therefore no spiral scroll is provided.

The bony fishes have a two-chambered heart with one auricle and one ventricle. These are further supported by sinus venosus and conus arteriosus. The red blood corpuscles are oval, biconvex and nucleated.

The olfactory sacs are dorsal and communicate with the mouth cavity only in those fishes that possess the breathing lung facility. The ear is a membrane and has three semi circular ducts. The lateral sense organs are well developed and are able to detect even minor vibrations and electric fields generated by nearby fishes.

The gonads are not recognizable outside. The reproductive ducts open outside and the fertilization thus is external.

Age

Mudgha Stage has witnessed the creation of a major group of aquatic animals that are collectively known as fish. A total of 15 or so major groups of fish were created during this stage. It is interesting to note that members of only five major groups have survived till today and the rest have become extinct. The Mudgha Stage, as evidenced from fossil records, started in Cambrian period and continued up to Triassic when it had peaked with highest diversity of species. The Devonian is regarded as the golden time of mudgha animals when the majority of fish groups were present. By the end of Devonian Period a transitory stage has started to equip animals for moving from aquatic environment to terrestrial environment. The real Mudgha Stage may therefore be a period of about 178 million years from the Cambrian (around 540 Mya) to the close of Devonian (around 362 Mya).

The primitive mudgha animals created around 540 Mya with chord-like outgrowth or stomochord and with gill slits are known as hemichordates. The fossils of animals possessing gill slits and V-shaped small blocks of muscles, an indication of the presence of small notochord as muscles are formed attached, are found in the deposits of Lower Cambrian (530 Mya). Fossils found in later deposits in the same site are of round mouthed jawless fish, cyclomates that were created by the close of Cambrian period. Two more types of jawless fish were created in the Ordovician period (500-438 Mya). Fossil records of Silurian times (438-408 Mya) have revealed the existence of at least six more groups of jawless fish (Southwood, R., *The Story of Life*, 2003).

Fossils found in China further reveal that jawed fish like placoderms were also created in early Silurian period. Their fossil finds were few in the beginning but the bedrocks of early Devonian period (408-362 Mya) point to their abundant presence. However, by the beginning of the next period they had become extinct. The other cartilaginous fish, Chondrichthyes, presumed to have been created in Silurian was later to be a common fish in Devonian. The Osteichthyes or the bony fish too were created in the Silurian but continued to live with only a few groups through Silurian and Devonian.

Devonian period is important in the sense that it witnessed the creation of the final product of Mudgha Stage in the form of a major group of lobe-finned, lunged fishes that flourished in large numbers. A subgroup of these fish called Latemeria or the Coelacanth is supposed to be the ancestor of all the tetrapods (amphibians, reptiles, birds and mammals). This group created in Devonian Period had reached its maximum diversity in about 140 million years by the Triassic Period (248- 206 Mya) (Southwood, R., 2003).

All the animals ranging from single-celled organisms created from clay around 3000 million years ago to osteichthyes formed at the close of Mudgha stage about 360 million years ago lived in water-bodies. They lived in every type of aquatic environment in oceans, lakes, rivers and other water-bodies. They occupied hydro-thermal vents or trenches in the ocean deeps or just floated on the surface water. They lived in muddy ocean-floor or moved around in shallow or deep waters. Most of these were created to survive in aerobic condition and needed oxygen for respiration. But while living in water they had been provided with special structures to get oxygen locked in water for their use. The light available there helped them in preparation of much needed food. Still some other water-living organisms were anaerobic in nature and survived in such ocean areas that were devoid of oxygen and light. They used sulphur and oxygen for respiration and heat, not light, as the source of energy in the volcanic vents in the deep ocean floors. Some chemoautotrophic bacteria that lead anaerobic life live in black sulphur layer deposited at places beneath marine sediments in the ocean floor. Apart from bacteria, some primitive animals without coelom like round worms, flat worms and jaw worms living in water lead anaerobic life.

Evolution of animals as changes in time is an example of God's Hand in Action as seen in answer to the following questions:

1. Who created gill slits and then made them to disappear in the adult animals at the close of Mudgha Stage?

2. Who shifted coelom to mid- dorsal position above the notochord

3. Who formed a non-ganglionated nerve cord and placed it above the notochord?

4. Who created a completely closed blood circulatory system?

5. Who located heart in a ventral position?

6. Who formed a well-developed cartilaginous vertebral column.

7. Who formed fins as appendages?

8. Who split the fins into fingers.

9. Who formed bony legs.

10. Who first created muscular lobes at the base of fins to help movement in muddy land surface?

11. Who created a gas filled swim bladder that in some fishes was also used as a lung for breathing.

12. Who created breathing sacs.

13. Who separately located urinogenital aperture away from anal aperture?

14. Who formed opisthonephric kidney for the removal of nitrogenous waste, urea?

He is God Who formed every minor functional part of the body

PART-II

Amphibious Life

(Stage V)

CHAPTER V

Amphibians

It appears strange to find that in a span of 2600 million years from the formation of first living cell up to the close of Mudgha Stage by the end of Devonian period, a wide variety of those animals were created that lived only in water. The land surface was totally neglected and not a single animal was found to have inhabited it during such a long period of life history.

Creation of fish however was not the goal of creation. The aim of God was to create man, His Vicegerent, together with other animals that were to live on land. Fish were only a part of the ladder in reaching out to achieve the goal. Then what prevented the appearance of Life on land for a very long period. It was in fact an inhabitable land and inhospitable atmosphere that prevented the appearance of life on land.

God was not unmindful of His aim to create man and therefore side by side of creating various life forms in water, He had started measures to make land surface suitable for habitation. The orogenic processes and vulcanism were activated to form various relief features like mountains, hills, plateaus, plains and valleys suitable for life on earth. Mountains were formed with deep roots to maintain proper balance on the surface and to prevent continuous shaking of the earth. Plains were made habitable and plants were grown as an easy source of sustenance for living organisms: "And We have set up firm mountains on earth..." (21: 31); " And He has set up on the earth mountains standing firm, lest it should shake with you; and streams and roads that you may find a way." (16: 15); "And the earth We have spread out (like a carpet); set thereon mountains firm and produced therein all kinds of things in due balance." (15: 19).

Another important change was to control unchecked ultraviolet radiation that had kept the temperature of the earth and its atmosphere high. It would often rise to such an extent that 'biologically suitable furnace' of the earth will turn into a 'blast furnace' that will destroy all living things. As compared to air, water has the quality of screening out harmful ultraviolet radiation and thus there was no danger from it to water dwelling animals.

The method adopted by God to control rising temperature was to use the same ultraviolet radiation that was a hindrance to life on land, as a tool to protect life from any harm. An electric discharge in oxygen or in air during a thunderstorm in the troposphere through photochemical reaction breaks oxygen (O_2) into two separate molecules ($O_2 \rightarrow O+O$). One of this oxygen atoms (O) is then combined with oxygen O_2 ($O+O_2 \rightarrow O_3$) and thus ozone (O_3) is formed. By a similar process ozone is transformed into oxygen: The collision of ozone (O_3) with mono atomic oxygen (O) transforms it through photochemical process into oxygen ($O_3+O \rightarrow O_2+O_2$). Thus, the action of ultraviolet radiation on ordinary oxygen by photochemical process constantly transforms oxygen to ozone and ozone to oxygen. The production and destruction of ozone is extremely slow but is a continuous natural process. Its production in the beginning may have outpaced its destruction so that gradually a mass of ozone layer was built around the earth in the troposphere. Although ozone is present at almost all altitudes, its bulk is concentrated as a layer in the stratosphere between 10 to 50 kilometres above the surface of the earth. Presence of ozone layer is good for the survival of animals and plants as it acts like a protective umbrella shielding all the organisms from exposure to dangerous ultraviolet solar radiation. It absorbs unwanted ultraviolet rays of solar radiation and allows only those radiation rays to reach the earth's surface that are essential for the maintenance and survival of life on land. Once the effective mass of ozone was built up in the atmosphere, the production and destruction of ozone was kept in balance. Even a slight change in the equilibrium level would have adversely affected life on the earth.

It was perhaps the time when ozone umbrella was perfected and land surface became habitable for the appearance of life, that God said, "...then We made out of that mudgha bones and clothed the bones with muscles..." (23: 14). This Verse clearly indicates that basic changes were being made by transforming cartilaginous soft material into hard bones and then bones were covered with muscles. It was an essential step when mudgha animals were being prepared to move from water onto land where bones will be needed to support their body weight and muscles would help them move or walk on land.

Fishes or other aquatic animals when in water are weightless as they are less influenced by gravity. It is because the specific gravity of protoplasm, the main constituent of the body, is approximately the same as that of sea water. The water-living animals therefore are able to float in water effortlessly. Even those fishes that are formed with the weight of partially formed cartilaginous/bony skeleton, do not feel their weight in sea as they either swim continuously or have been provided with a swim bladder to compensate the weight and thereby continue effortlessly to float or keep balance in water at any depth.

But if same aquatic animals are brought out of water on to the land surface they are bound to be affected by the gravity of earth. Their weight will increase and they will need bones to hold various parts of their body in position otherwise the body will collapse. The thickness of the bones will depend on the size of the body. Large bodies on land will have greater weight and will consequently require thick and large bones to bear the body weight.

Mere creation of bones is immaterial if muscles are not attached to bones. Muscles are required to support skeleton for a proper posture of the body and for movement of its parts. More importantly muscles help create movement of bones for walking and running and for various other actions of the body parts. The contraction and pull of the muscles start the movement of bones. Bones on the other hand act as an anchor for the effective pull and contraction of the muscles. The two, bones and muscles, are interdependent and function properly only if they work together. It was therefore necessary that if the animals are made to dwell on land, they should be provided with weight bearing bones covered with muscles for perfect functioning of the body. Moreover, land dwelling animals consume 40 per cent of their energy just for carrying their weight. Thus God, while introducing modifications to suit smooth transfer from sea to land, created bones (skeleton) and developed new muscles to meet out any deficiency of energy.

Other Requirements of Land Dwellers

The physiology and anatomy of land-dwelling animals is totally different from those living in sea. The transition from water to land requires complicated modifications in the animals to make them functional on land surface. God had laid down the foundation by providing bones and muscles, the basic and necessary requirement, for oncoming changes in the organ systems of the animals living in water. Still drastic changes and improvements were required to make these functions suitable for terrestrial life. Some of the important changes carried out in the skeleton, circulatory system and excretory system are mentioned here.

Skeleton

It has been pointed out earlier that as compared to sea, the body weight of animals on land is increased owing to gravitational pull of the land. In order to bear the weight, it is essential that land animals be provided with strong supporting legs. This was done by God in two ways: Firstly, the fins were split into fingers to distribute the weight at different places rather than it be concentrated at one point. Secondly a complete weight bearing bone lever system was created to join fingers with the vertebral column. The half-formed bones in the fins were converted into full bone and joined to the vertebral column to make it a fully formed supporting leg. Of Mudgha fishes the first to invade land is supposed to be *Latimeria*, the lobe-finned fish of a subgroup Sarcopterygii. The muscular lobes were created and strengthened at the base of anterior dorsal fin to help them powerfully thrash the mud on land and move forward. They could not go far into land as their body supporting bone was partially formed. This fact has already been mentioned by God that at the time animals were being made ready to bear the weight on land; "...then We made out of that lump of flesh (mudgha) bones and clothed the bones with flesh (muscles)..." (23: 14). It becomes more obvious by comparing the structure of Pander ichthys which supposedly is the ancestor of land animals with that of the first imperfect amphibian. The fins of the ancestral fish were split into fingers and supporting bony legs were formed.

The leg bones were still imperfectly formed and therefore could not properly support the weight of the body. This shortcoming was further improved in early amphibians, Acanthostega, that were created with fully formed leg bones joined with the backbone. The bony legs together with fully formed arms helped them move on land surface too.

Beside the formation of full leg bone, the individual vertebrae that were cartilaginous so far were converted into bones. All the vertebrae were joined together to fit into one another and become a fully formed backbone. It completed an important part of the load bearing structure, and the skeleton of amphibians was ready to carry the weight of the body.

Circulatory System

Blood had already been formed in aquatic animals and moved in closed circulatory system. It conveyed respiratory gases to and from gills and was also a medium for transport of nutrients and water products. The function was modified when trachea was formed in animals as a means of survival on land. Henceforth the blood in land dwelling insects and animals continued to transport nutrients and water products, but its function as carrier of respiratory gases was stopped and taken over by trachea.

Excretory System

Animals need energy for their metabolic activity and the process creates an end product or waste. The waste needs to be thrown out quickly in order to save body from harmful effects of the toxic material. Water dwelling animals mainly create nitrogenous waste, ammonia that in most cases is diffused through body surface or through excretory pores. The annelids excrete ammonia through coiled tubes, nephridia, while arthropods do it through antennary or green glands that directly open to the exterior. Molluscs are created with one or two pairs of sac-like kidneys that excrete ammonia or uric acid through a direct opening into the mantle. Echinoderms diffuse ammonia through gills or tube feet. Sac-like kidneys created in mollusks were improved upon in fish that were formed with opisthonephric kidneys for the removal of nitrogenous waste—urea or ammonia.

Land dwelling animals, however, needed special machinery for flushing out more toxic urea or uric acid because any delay in it would lead to flow of urea in the blood that would cause damage to other organs. To meet out this exigency, God created animals with metanephric kidney with attached urinary bladder to remove urea from the system. These were transitory animals leading a dual life in water as well as on land. Later on, the metanephric kidney attached with a bladder was further improved and animals were created with a perfectly formed bean-shaped kidney so that they could dwell wholly on land surface.

It is thus clear that the aquatic and terrestrial life is different and needs substantial changes in anatomy and physiology of animals switching over from water to land. All such animals are classified by Zoologists as Tetrapods. They are supposed to have descended from lobe-finned, lunged fish called Latemeria or the Coelacanth formed by the end of Mudgha Stage. These fish with a muscular lobe attached at the base of the fin had started to invade on to land by thrashing their way in muddy waters.

The word tetrapods stand for four legged animals but in fact it includes all the limbed animals whether carrying two legs or four legs. Tetrapods are divided into two groups: (1) Amphibians that lead a dual life in water and on land like present day frogs, toads, salamanders and newts, and (2) Amniotes that are land and water-dwelling animals like reptiles and birds.

Amphibians

The transitional phase when God made bones and clothed them with muscles and introduced improvements in other organ systems to arm animals with such tools that helped them live in water and encroach upon land too is known by scientists as the phase of amphibians or the phase of animals leading a dual life. Amphibians were created as a link between the two with elementary changes to make them capable of leading a dual life. Such animals breed in water and lead their fish like larval life called tadpole in water, swimming with tail and respiring with gills. When adult, they come over to land, use limbs for movement, and respire through lungs and skins as well as through buccopharyngeal lining in the mouth. Their whole life passes through the process of metamorphosis as they start as a tadpole in water and end as four-legged air-breathing adult on land.

God has created all living things from water (2: 30). As such, water is essential for metabolism and every living organism, whether in water or on land, is dependent on it for its development and growth. Amphibians at the tadpole stage face no problem as they remain in water, but once out on land as adults, they need water intake to satisfy metabolic need of the body. Moreover, while moving on land in sunlight, their body needs some protective cover to prevent water loss through evaporation. The other land-dwelling animals created later had a skin as a safeguard against this eventuality. But the smooth skin of water dwelling animals, without scales, is not fully prepared for it. A beginning was certainly made in amphibians to convert skin to be a little evaporation proof by making it slimy through the presence of mucous glands. Thus, they could live in wet environment and keep their skin moist. It is yet another measure to minimize evaporation of body water.

The endoskeleton in amphibians is largely bony consisting of vertebral column, skull and limbs. The first vertebrae or atlas (in Greek mythology god Atlas is supposed to hold the earth on his shoulders) is formed to facilitate movement of the head. Skull is dicondylic, hence it cannot move sideways. There are two pairs of jointed pentadactyl limbs. In some, like frogs and toads, the hind limbs are larger than the fore limbs, while in some, like in salamanders and newts, both are of the same size. There are 4 to 5 or fewer digits in each limb but they are devoid of any claws, nails or hoofs. The fossil of probably first primitive amphibian, Acanthostega, shows the formation of eight toes in the forelimbs and seven toes in the hind limbs. Pentadactyl limbs with 5 fingers are found throughout vertebrata though fingers may secondarily be lost to make a single digit functional toe like that in horses.

Amphibians are the first animals created with protractible muscular tongue. Fish too had tongue but it was non-muscular and non-protractible. The tongue and bucco-pharyngeal cavity are lined with epithelium that produces sticky mucus. The mouth is large and possesses acrodont teeth in upper or in both jaws. Teeth generally have wavy lines caused by in-folding of the enamel surface and owing to this typical feature these four-legged vertebrates leading a dual life are also called labyrinthodonts. The alimentary canal on the other end opens into cloaca for throwing out excretory products.

The respiration in amphibians takes place in various ways. Larvae use gills for respiration but in adults the function is performed by lungs. Respiratory gases enter into external nares and moving through internal nares reach pharyngeal region. Skin and lining of buccopharyngeal cavity are also used for exchange of gases.

Heart has three chambers, an improvement over fish that have a two-chambered heart. Lungs supply cleaned blood to left and right auricle but the ventricle pumps out mixed blood. The circulatory system is much developed.

Kidneys formed in amphibians are pronephric in larvae while in adults they are mesonephric or opisthonephric. Urinary bladder is formed for the first time in amphibians. Urea is excreted by tailless variety and ammonia is excreted by larvae and tailed forms.

Sense organs are present in the form of eyes and ears. Eyes are protected usually by three moveable eyelids; one is transparent nictitating membrane, remnants of which are found in higher vertebrates. A middle ear is also formed and is connected with internal ear. There is no external ear, instead the opening of middle ear on the surface is covered by tympanums that receive sound waves and thus animals can hear.

On the whole, amphibians were created with certain advancements over their immediate predecessors, fishes. All the improvements were helpful to them in leading a terrestrial life. The advancements are: 1) Two pairs of jointed pentadactyl limbs for locomotion, 2) a protractible true tongue to capture food, 3) formation of lungs for pulmonary respiration, 4) presence of 3-chambered heart (with two auricles) for receiving and pumping both arterial and venous blood, 5) creation of metanephric kidneys and urinary bladder for flushing out urea, 6) three eyelids and tear glands for protecting eyes from dust, and 7) a middle ear to transmit aerial sound waves to the internal ear.

Age

The fossil records reveal that the first four-legged amphibian, probably Acanthostega, was created in the late Devonian Period. A large number of fossils of amphibians were identified to exist in the last 10 million years of Devonian period from various parts of the world. Later on, however there is a gap of 20 million years in the fossil records of amphibians. Maybe it was due to tectonic upheavals and large-scale earth movements of the day that prevented the formation of fossils or destroyed them all. After the gap, a wide variety of amphibians known as labyrinthodonts appear in the fossil records around 340 Mya in the Carboniferous period. "...the labyrinthodonts lasted through the arid Triassic Period and one (Koolasuchus) has been found in Australia from the fossil records in the Lower Cretaceous. Modern amphibian groups are not known from the fossil record until the late Triassic, though they must have evolved earlier from a different group of amphibians found in the Carboniferous." (R. Southwood, 2003).

Answer to the following questions lies between God and Evolution?

1. Who first provided muscle power to arms and legs?

2. Who made a perfect supporting leg by joining it to vertebral column?

3. Who made two pairs of jointed pentadactyl limbs?

4. Who fully converted individual vertebrae into bones and joined them to make a complete backbone.

5. Who facilitated amphibians to respire through lungs and skins as well as through buccopharyngeal lining in the mouth?

6. Who created opisthonephric kidneys?

7. Who for the first time formed urinary bladder?

8. Who first created eyelids and tear glands for protecting eyes from dust?

9. Who first made the skin of amphibians slimy by the presence of mucous glands?

10. Who first created a protractible tongue?

11. Who first formed a middle ear to transmit aerial sound waves to internal ear?

12. Who created a three chambered heart?

13. Who first created such animals that could lead a dual life in water as well as on land?

A switch over from life in water to life on land required drastic changes that only Supreme Power could achieve.

PART-III
Life on Land
(Stage VI)

CHAPTER VI

Terrestrial Life Forms

The lifestyle of land-dwelling animals is quite different from those that live in water because they belong to two totally different environments. Water dwelling animals are generally sessile, i.e., they can obtain their food by just waiting or staying at one place, while land dwellers had to be mobile and go out searching for food. It is because the primary source of food in water, the plankton, is available abundantly at or near the surface of water and moves with it to reach the feeder. The availability of plankton at the surface is the outcome of typical nature of water. Water filters light so that its intensity decreases with depth and finally becomes zero at certain point. Ocean depths therefore are dark while the surface water is full of light. The photosynthesizing bacteria and protists use sunlight and make primary food through the process of photosynthesis at the surface. The food supply therefore is abundant at or near the surface of water and is consumed by sessile animals.

On the other hand, plants, the primary producers of food on land, are rooted to the soil and do not move to provide food to animals as is done by water floaters. Instead, the land dwellers have to be mobile and go to the plants or trees to satisfy their food requirements. Light on land surface is available everywhere and photosynthesis or the process of primary formation of food by plants is possible all over the land. The difference in food production by plants lies in the competition for getting as much sunlight energy as they can. Larger trees capture greater amount of light and prepare greater quantities of food. Taller trees prevent light from reaching the ground and hinder the growth of plants in their shadow. The plant food therefore is not available abundantly everywhere. Its presence is variable on the ground surface as well as high above it consonant with the height of the plant. In view of varying availability of food supply, animals were created with varying physical constitution to facilitate them procure plant food from the ground.

Another point of difference between water-dwellers and land-dwellers is in the fact that most of the groups of multicellular animals with different body organizations were created in water-bodies while only a few were formed on the crust. There are 37 groups or Phyla of which 34 are purely water-dwellers while only 3 are found on land. The formation of 34 major body changes in water living organisms took a long period of about 750 million years since the first multicellular organisms were formed. As against it, land dwelling animals took only 350 million years until third major body change took place to form last and final product, the man.

The low number of terrestrial Phyla was nevertheless compensated by creating innumerable species in different classes of land Phyla. Species variation was made possible by the presence of a large number of micro-habitats on the land because of micro level changes in temperature, rainfall and soil. Water, on the other hand, with its high specific gravity of heat, is subject to small seasonal and regional variations in temperature. Small variations in watery ecosystems do not form diversity in species, hence their number in oceans is restricted. As a result of these differences, the land fauna is most varied both in form and colour, and is spectacular and attractive.

To cross the barrier between water and land God created a link between the two in the form of amphibians that lead a dual life of living in water as well as on land. But these animals could not break away from water as they were dependent on water for spawning. The dependence on water for embryonic development was a great hurdle in the transfer of organisms over to land and to be capable of pursuing a lifestyle disconnected from water-bodies. God solved the problem in one stroke by providing necessary pool of sea water within a water proof protective cover on land itself for use by the organism when it was a foetus. It was achieved by creating a hard shell enveloping the egg that protected inside water from evaporation. The eggs of amphibians have jelly like structure with a porous covering that could easily be desiccated in the sun or on dry land. In contrast, the hard shell protected the egg from drying and kept the watery environment safe inside for use by the embryo on land, away from any water body (Fig. 5). It was a revolutionary giant leap in the history of creation that completely altered the style of living from inside water to

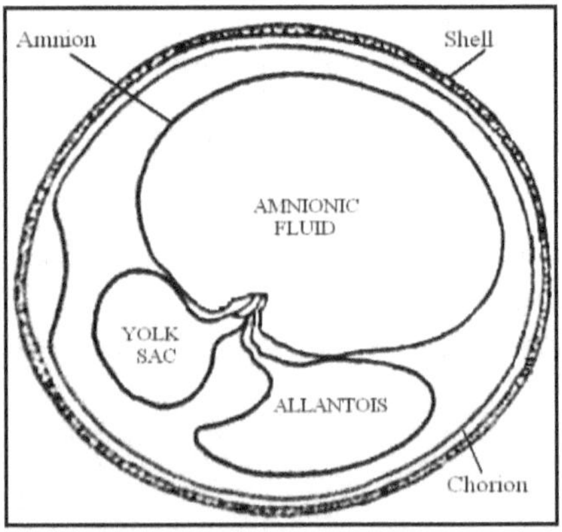

Fig. 5: Constitution of an Egg

that on the land surface. Eggs with the shell are laid on land and are the first brick in the great life structure built on land.

Within the shell, eggs have all the essential items required for internal fertilization and the development of an offspring. Water is present in the egg enclosed by an inside membrane called amnion. It has almost the same composition as that in the sea and the embryo floats in it with the feeling of affinity as if being developed in the sea. The embryo is connected to membranous sacs; one is the yolk sac that provides nutrition to the developing foetus, the other is gas filled allantois that contains much needed gas for respiration and then there is a cavity that provides space for accumulation of nitrogenous waste. As the water inside is limited, the waste product is uric acid and not urea. All the three sacs--- amnion, the yolk sac and allantois--are enclosed by an outer foetal membrane that is known as chorion. The shell is water-proof but is permeable to gases. The egg contains all the necessary material needed by the growing foetus and in the quantity required for the period until the embryo comes out.

The formation of a shelled egg facilitated the creation of two major Phyla of land- dwelling animals, i.e., four-legged reptiles and feathered birds.

Reptiles

Reptiles were perhaps the first animals to be created in a terrestrial environment. They laid shelled eggs and were thus able to continue breeding on dry land. Another protective cover in the form of epidermal scales and dry skin helped them in survival on land by checking the loss of water through evaporation from the body. Moreover, they had also been provided a mechanism of internal fertilization because gametes could not have survived on dry land. They also had claws for attack and defence.

The reptiles had two pairs of pentadactyl limbs that possess 5-digits. The limbs are directed outwards so that they were able to creep on the surface. Some reptiles like snakes and a few lizards are formed without limbs which are secondarily degenerated. They are able to creep with the help of ventral scales and ribs using the lateral body wall muscles. The body is divided into four segments: head, neck, trunk and tail. The head is joined with two specifically formed vertebrae that permit it to move independently of the other body parts. The mouth is large and both the jaws have teeth, Turtles and tortoises, however, have no teeth, but have horny beaks instead. Beside mouth the digestive system has an alimentary canal that opens on the other side into cloaca. The waste material is excreted by a pair of metanephric kidneys mostly in the form of uric acid in those living on land, but in turtle and alligators it is in the form of urea and in crocodiles, it is ammonia. Urinary bladder is present in almost all the reptiles except in snakes. Respiration through lungs is made efficient because of rib movement that helps the expansion and contraction of the trunk.

The circulatory system is an improvement over that of amphibians in that the heart is 4-chambered with 2 auricles and partly divided 2 ventricles. Amphibians had one ventricle which pumped out mixed blood. In reptiles the ventricle is partly divided so that the oxygenated blood from the lungs and the deoxygenated blood from the rest of the body do not mix completely. Crocodiles however have a fully formed 4-chambered heart for blood circulation.

The body temperature in reptiles is variable as it depends on heat available in the environment. Unlike birds and mammals that keep constant blood pressure, reptiles are regarded as cold blooded that may undergo hibernation in severe cold conditions. These animals try to keep their body temperature higher than the surroundings by exposing themselves in the sun for some time. In the mornings particularly, they warm up before they become fit for any energy consuming activity.

It is generally believed that reptiles were created by incorporating certain improvements over salamander, a land-living amphibian. Reptiles are terrestrial and are created with a hard shell and a fluid-filled embryonic membrane, amnion to successfully breed on dry land. They possess horny scales made of keratin to check evaporation of body water, claws for defence, and two atlas vertebrae under head to make it firm. The two atlas vertebrae do not allow head to move sideways as compared to birds that possess single atlas vertebra that allows the head to turn sideways 180°. Beside two auricles, the ventricle of the heart is partly divided into two chambers to keep the oxygenated and deoxygenated blood separate. The ribs have been formed to help rapid expansion and contraction of lungs and thereby increase the efficiency of pulmonary respiration. They are also provided with a pair of metanephric kidneys.

Classification of Class Reptilia

Class Reptilia is divided into three subclasses on the basis of presence or absence of temporal fossae or holes behind the eye. The Subclass without any hole is called Anaspida, those with one hole are Synapsida, and those having two holes, one on each side, are known as Diapsida.

Anaspida is represented by the green turtle, the tortoise, and the terrapin. All anapsida are generally herbivores feeding on plants.

Synapsida are most advanced reptiles that possess mammal-like features of 4-chambered heart, a diaphragm, secondary palate, spongy lungs, and thecodont teeth. The examples of synapsid are crocodile, alligator, and gharial. All the synapsida are amphibious reptiles and are carnivorous in habit.

The existing members of diapsids are tuatara, lizards and snakes. Tuatara is a carnivorous reptile. Lizards are generally insectivores but some as spiny-tailed lizards are herbivores. Examples of lizards are wall lizard, girgit, Indian monitor or goya, garden lizard, spiny-tailed lizard, flying lizard, limbless lizard, etc. The most important examples of snakes are cobra and ajgar (python).

Reptiles of huge size were also created that roamed on the earth for about 150 million years, but suddenly disappeared and are now extinct. They are known as terrible lizards or dinosaurs. The extinct dinosaurs were of two types: reptile hipped and bird hipped. The reptile hipped thunder lizard and plated lizard were herbivores, and tyrant lizards are carnivorous. The bird hipped reptiles or small sized ornithischian dinosaurs are supposed to be the ancestors of a separate line of birds.

Age

Reptiles appeared on the earth in the beginning of Carboniferous Period about 300 million years ago (362-295 Mya). The earliest reptile in fossil records is that of 'Lizzie', a lizard like reptile found in the rocks of early Carboniferous Period in West Lothian, Scotland. All types of reptiles were formed by the Cambrian times (295-248 Mya), and were widely spread in the Mesozoic Era. They were dominant animals in th Jurassic Period (206-148 Mya) and therefore that period is also known as the age of reptiles. Moreover, dinosaurs that appeared as small size animals in Triassic period (248-206 Mya) had grown into large sized reptiles in Jurassic Period. Dinosaurs however could not live for long and suddenly disappeared from the scene or became extinct by Cretaceous times (145-65 Mya).

Birds

Birds are tetrapod vertibrates that are covered with feathers and can fly in the air. Their fossil records point, albeit with some disagreement, that birds have been formed with modified improvements from reptiles, especially flying dinosaurs. Apparently the two, birds and reptiles, stand wide apart in their anatomy, structure, and metabolism and do not seem to have any ancestral relationship. The constitution of the two is made to function differently. One walks on the land while the other also flies in the air. Some species of both however, are similar in that they can, beside their domain, also swim in water.

Birds and reptiles both have two pairs of limbs but their shape and functions are different. The forelimbs in the birds have been modified into wings used for flight. The reptiles use all the four limbs simultaneously in a synchronised way for movement, but the birds use only two limbs at a time: forelimbs for flying and hind-limbs as body support for resting or perching and for walking, hopping, wading or swimming.

The endoskeleton in reptiles is composed of thick, solid, weight-bearing bones, but the same in birds are delicate, hollow and light to permit easy air lift of the body. Similarly, the exoskeleton scales of reptiles have been changed into epidermal feathers in the birds. The scales are hard layers formed as surface cover on the skin, but the feathers punch the skin of birds and are nailed therein like pegs. The birds also have scales on their feet. Another difference in the endoskeleton is seen in the formation of pelvic girdle. The birds are made with a special enlarged pelvic girdle to help their bipedal movement, keeping their vertical stance and providing support while resting on two legs. Small girdle suits only quadruped reptiles that rest horizontally on the ground. The birds also possess an enlarged sternum with a median keel for the attachment of flight muscles.

The birds have been formed with a unique mechanism of respiration that is very different from other animals. The lizards and other land-dwelling animals inhale air through trachea and exhale it through the same passage. In birds fresh air passes through bronchial tube, reaches tiny air sacs in the lungs where oxygen and carbon dioxide exchange take place and the stale air comes out through the same passage. The respiration in birds is a complicated process. It is mainly accomplished through 7 to 9 air sacs attached to the lung. Two inhalations of fresh air are required to complete the respiration cycle. The air moves in the lung by pressure changes in the air sacs. Fresh air passes over the gas exchange surfaces of the lung during both inhalation and exhalation. This mechanism provides a constant supply of fresh air and the lungs are able to continuously perform gas exchange process.

Another major change was done in the design and function of lungs between birds and other mammals. The large lungs of other animals inflate and deflate with successive inhalation and exhalation of air. In birds, however, the lungs are narrow and rigid. They do not expand or contract like the lungs of mammals. Tiny air sacs inside the lungs cannot fill with air and empty again. If ever the lungs are deflated, they cannot be inflated easily, thus leading to fatal consequences. It is necessary for the survival of birds that lungs should constantly remain filled with air flow, thereby preventing any possibility of deflation. A continuous unidirectional flow of fresh air in the lungs made possible by unique mechanism of air circulation in birds fulfils their high energy requirement.

Yet another vast difference relates to body temperature between cold blooded reptiles and warm-blooded birds. Reptiles have very slow rate of metabolism and cannot regulate their own temperature. They depend on the external source of energy from the environment and therefore have to warm up in the sunlight before they can perform any physical activity. The birds in contrast are homoiothermic or have the power to regulate their body temperature to keep it high in the range of around 38-41°C. They have high metabolism rate that helps in raising the temperature quickly. Birds being endothermic consume a lot of energy to maintain body temperature and therefore need to feed more often than the reptiles.

Beside major alterations, some other improvements over reptiles have been carried out in birds. Instead of partially divided ventricle in lizards, the heart in birds has fully formed 4 chambers with two auricles and two ventricles. The oxygenated and deoxygenated bloods remain separated from each other as a result of two ventricles. The heart in birds is relatively large and fast beating. It helps in quick supply of blood during flight. The birds have larger brain with respect to their body size, a well-developed eye sight, and better power of hearing than the reptiles.

Classification

The birds belong to Class Aves which is divided into two Subclasses. Archaeornithes, a primitive extinct bird possessing both reptilian and avian features with homodont teeth, long tapering tail, weak and keel-less sternum, non-pneumatic bones, and claws in fingers. The ancient lizard bird is an example of this Subclass. The other Subclass Neornithes has some existing and some extinct birds that have no teeth, a very short tail, possess keeled sternum, pneumatic or hollow bones, and unarmed fingers. The examples of such birds are of those that cannot fly like ostrich and kivi, those that can also live in water like penguins and duck, and those that can fly like house sparrow, pigeon, fowl, house crow, koel, parrot- parakeet, peacock or pea fowl, owl, vulture, kite, etc.

Age

It is generally agreed that birds have originated from dinosaurs, so much so that they are also called 'feathered dinosaurs' or 'glorified reptiles'. The fossil records indicate that dinosaurs with bird-like features first appeared in Late Triassic and were widely present in Jurassic and Cretaceous periods. Full-fledged birds appear in the fossil records in the Jurassic. The period of about 36 million years from Late Triassic to Jurassic (237-201 Mya) is geologically a short transitional time to introduce modifications. It required many major alterations in lizard, like formation of feathered wing, unique mechanism of respiration, and capacity to regulate temperature. The design of creation by God runs its own course and changes are incorporated one after the other. But a lot of diverse modifications achieved in a short geological span have hindered scientists to ascertain the actual sequence of modifications. Fossils in a short period are a mixed lot and fail to present a clear picture about what course of action was followed in the creation of birds.

Early biologists believed that in the first instance pelvic girth was changed and enlarged to be like that of the birds. They therefore divided the dinosaurs on the basis of hip structure into two groups; Ornithischian or bird-hipped, and Saurischian or lizard hipped. Later researches however indicated that the name bird-hipped was wrongly applied to them as they are not the fore-runners of birds. The elongated hip girdle of course is like birds, but they are quite different from them in other anatomical features. Ornithischians have short legs, giant body, scale covered skin, and are quadruped herbivores. These features are dissimilar to birds and may not have helped the least in their formation. Contrarily, Saurischians with long hind legs, short forelegs, thin structure, skin often feather covered, and bipedal carnivorous habit are quite close to birds.

There are nevertheless two dissimilar types of Saurischians. One, Sauopomorpha has members that are large size, quadruped herbivores. The other, Therapoda has bipedal carnivorous members that range in size from less than a metre to about 14 metres. It was one of the Therapoda dinosaurs, the pterosaurs, that were actually the ancestors of birds, as they were first to be formed with membranous wings in the Late Triassic period.

Some species of pterosaurs were created with a membrane that joined the hind limb with the fore limb. The membrane made a bat-like wing and the pterosaurs began to jump long distances from branch to branch or fly in the air with the help of the membranous wing. It was also the time when most of the theropoda dinosaurs were covered with feathers. It was only when a tubular structured quill was created that great strides were possible towards flight. The quills were hooked to fore limbs in specially formed complex design so that they made a rigid entity and birds were created to fly with feathered wings. It cannot be said for sure whether membranous wing started the flight or feathered wing did it, or the two were created simultaneously but independently of each other. Apart from what course was undertaken by God in their formation and how they took flight, the fact is that the Cretaceous Period remarkably shows the presence of their several groups.

Mammals

Appearance of reptiles on land, away from sea water, was made possible through the mechanism of sea-like water provided inside the hard-shelled eggs. It also gave rise to other flying life forms that continued their progeny through hard-shelled eggs.

Side by side, and before the formation of egg-laying birds, another mechanism was created by God wherein the sea water conditions were provided not in an egg but in a bucket inside the body wherein young ones developed and were delivered. The animals thus formed were drastically different from their ancestral reptiles. The sex organs in them were formed separated from each other. The fertilization took place inside the body, not like reptiles where it occurs inside the egg. The development of the new form was directly inside the female. The foetus was connected to the uterus with a line called placenta through which it could receive nourishment and oxygen from the mother's body. When fully developed the young ones were delivered but had to still depend for survival on the nourishment provided by the mother in the form of milk. The suckling continued until the body grew enough to accept outside food. The nourishing milk comes out through the mammary glands. The creatures born suckling from the mammary glands are identified by this facility and are called mammals.

Another major change introduced in the formation of mammals is that they are warm blooded or homoiothermic as compared to cold blooded reptiles. Being warm blooded, they have a high rate of metabolism and are capable of performing different types of activities easily as compared to reptiles. They do not depend on external heat for making them active as is necessary for reptiles but are able to regulate their body temperature themselves. If they are living in cold areas their skin has been provided with hairs or furs to prevent the loss of body heat and to keep them warm. As against it, those living in warm areas have been provided with sweat glands to release excessive heat and keep their body temperature normal.

Besides giving live birth and homoeothermic nature, mammals are created with other physiological, anatomical and structural changes that have made their body functions more efficient and advanced as compared to their ancestral reptiles. Their two pairs of pentadactyle jointed appendages or the forelimbs and hind limbs are specially formed for walking, running, climbing or jumping. The neck is formed with seven vertebrae of which the two are specially made to help the movement of head. The lower jaw is formed of a single strong bone with teeth on both sides. A secondary plate had been added to separate the digestive tract from the respiratory tract. Moreover, the opening of trachea or glottis is covered with a flap like structure called epiglottis. It prevents the food from entering into trachea during feeding. The mouth is relatively small in comparison to reptiles and possesses moveable lips.

There are teeth in both the jaws that are heterodonts or of different types like premolars, molars, incisors and canines. Their presence varies in different classes of mammals. Some possess one or two types while some others may be provided with all the four types of teeth. The most important improvement in the mouth of the mammals is the provision of two salivary glands that start the digestion of food within the buccal cavity itself. The head is provided with an ear that has three parts: the external or pinna that collects sound waves and middle and inner ear. The eyes possess moveable upper and lower eyelids.

Internal organs in mammals are placed in four compartments of the coelom. The heart is placed in one pericardial cavity; the lungs in right and left pleural cavities, and the abdominal cavity contains the viscera. A special feature, however, is the formation of a muscular, transverse and circular partition known as diaphragm that separates the thoracic cavity housing the heart and lungs from the abdominal cavity. The presence of diaphragm increases breathing efficiency by helping proper inspiration and expiration.

Circulatory system is controlled by 4-chambered heart. They help in maintaining two types of blood circulating separately in the body. The oxygenated blood reaches every part of the body and similarly the deoxygenated blood is collected from every extremity.

Kidneys in mammals are metanephric and bean-shaped. The urine is collected in urinary bladder from where it is excreted as a fluid.

Classification

The nature of gradual transformation from reptiles to mammals may still be eluding palaeontologists, but the mammals occupying land today themselves tell the story as to how it had happened. Mammals are recognized by their intrauterine development, direct birth of young ones, and nursing through their mammary glands. The nature of this major change from egg-born reptiles to directly young-born mammals form the basis of scientific classification of living mammals like Prototheria and Theria.

Subclass Prototheria

Prototheria are more primitive animals bearing features of both reptiles and mammals. Their representatives are duck-billed platypus and spiny ant-eater found in Australia, Tasmania and New Guinea. These primitive mammals also carry reptilian features. They are egg-laying, develop inside the large, hard shelled egg during incubation and are born naked and blind. They are suckled like mammals but their mammary glands are not fully formed as they are devoid of teats or nipples. The other reptilian features that they still maintain are one opening or cloaca, and a reptilian ear in the form of just a hole. Their body temperature like reptiles is low, lying between 25-28°C. They are however covered with hairs like mammals. Their hands though small are strong enough to provide separate vertical stance like mammals.

Subclass Theria

Theria are real mammals that develop in the uterus and give birth to the young ones directly. They have no cloaca, instead, the digestive tract and the urinogenital duct has separate openings outside the body. The subclass Theria is divided into two infraclasses on the basis of yet another major change incorporated in their creation. The early formations created with small placenta and small developing period are classified as Metatherian and those formed subsequently with large placenta and long period of development are classified as Eutheria.

Metatherians have small, simple placenta owing to which their intrauterine development is brief and youngsters are delivered in an immature, helpless condition. They are therefore kept for some time for proper growth in a special pouch formed outside the abdomen of the mother. This pouch is called marsupium and the metatherians are thus also known as marsupials or pouched mammals. The new born is nursed by the mother through mammary glands that open into the marsupium. The anus and the urinogenital tract have separate openings but both are controlled by one common sphincter. Metatherians unlike toothless prototherians are created with teeth but they are monophyodonts. Their body temperature is moderately warm ranging between 32-38°C.

Eutherians are true mammals that are born fully developed. They have a large placenta and a long gestation period that helps full development of the young ones inside by the time they are delivered. They need not be put in any pouch like marsupials for further growth. Since they are fully formed while still in the placenta, they are known as placental mammals. The digestive tract and urinogenital duct in eutherians have separate apertures controlled by separate sphincters. They have diphycodont teeth. Eutherians are warm blooded animals and their body temperature ranges between 35 and 40° C.

On the basis of broad dissimilarities among eutherians, they are arranged into four groups: 1. Most primitive insect-eating placental mammals like musk shrew (Hindi: *chhachhunder*, hedgehog, and small and large bats. 2. The gnawing canine less placentals like rodents and rabbits; 3. Canine bearing gregarious placentals like carnivores, ungulates, and elephants; 4. Aquatic placentals like whales, dolphins and porpoises.

Age

Of the three major groups of reptiles, there was one, synapsids having one hole behind the eye that had mammal like features. Such mammal-like reptiles are supposed to have been developed as true mammals. Creation of mammals is a crucial step in the development of life forms as it provides the base for ladder leading to ultimate creation of man.

Early mammals were herbivorous and their dicynodont members were present abundantly in the Permian times. By the close of Permian times around 248 million years ago, a catastrophe known as Permian extinction destroyed most of the reptiles and mammals. It will be of interest at this point to note that whenever in Qur'an God mentions creation with His own Hands (36: 71), invariably that means 'creation using power. The proof of intense radiation created by the impact of a huge meteor is found in layers of radioactive material in rocks formed during this time. The impact of huge meteor leading to radiation is regarded the cause of disappearance of dinosaurs in the Cretaceous Period and it also is the time heralding the creation of mammals.

One species of herbivorous group Lystrosaurus, however was less affected by the catastrophe and was abundantly distributed in the Early Triassic. A number of species of dicynodonts are found in the fossil records of Middle and Upper Triassic but all of them had died out at or near the end of the Triassic.

In late Permian one group Therocephalia started eating insects and by the beginning of Triassic became full-fledged insectivores. It was perhaps from this group that cynodonts were formed in the Late Permian. Cynodonts were abundant in the Triassic. Initially they were insectivores but gradually became diversified to become herbivores and carnivores. They also exhibited some clear mammalian features (Southwood, R., 2003).

The introduction of mammalian features in the reptiles like in other earlier formations was not sudden. True to the way of God, one feature may have been formed in a species, and then developed into a perfect function in the descendant species. The other mammalian features may also have been formed this way. Paleontologists have so far not been able to find links in the phylogeny of mammals. There are no fossil evidences to show how reptiles changed into mammals. There is no evidence to show how cold-blooded, egg-laying reptiles covered with scales turned into warm blooded, fur or hair covered mammals that gave live births and suckled their young ones. Without any fossil records to prove gradual transformation, mammals seem to appear suddenly on the surface of the earth. They did not only appear suddenly, but appeared in all their different forms and Orders like mice, horses, primates, bats, etc. The fault however does not lie on the Creator that he broke the rule and created fully-formed various types of mammals suddenly. It is the weakness of the scientists that they have not yet been able to find the links showing gradual improvement upon reptiles towards their transformation. The blame should not totally lie on scientists as the change from reptiles to mammals was taking place at a time when earth and its environment was passing through great upheavals and the survival of mammals and their fossilization was a difficult task.

The earth has witnessed three major and some minor extinction in fossils history from Late Permian around 256 Mya when mammalian features began to be formed. Whatever causes may have led to such extinctions, and scientists have suggested many, the fact remains that they have obliterated links of gradual formation of mammalian features. Some sporadic findings and their microbiological study helped us guess how the mammalian features could have been formed.

The Cynodonts, a group of reptiles, were the first to be created with some of the mammalian features: "…The We made out of that Mudgha, bones and clothed the bones with muscles…" (23: 14). The bones of their fore-limbs were formed first. The bones were strengthened with muscles and they started to move with a mammal like erect anterior posture. The vertical stance helped them to move with greater speed and with enhanced and efficient breathing. The bones were further strengthened later on and true mammals were created that were able to move standing on all four limbs or then onwards with only two legs. Cynodonts were formed with a set of differentiated teeth like incisors, canines and molars used for chewing. The other mammalian feature that they were created with was diaphragm that facilitated efficient breathing side by side their body movement. The cynodonts however became extinct by the end of Triassic when, as seen from the fossils, full-fledged mammals were formed.

A recent fossil finds of ancient mammal sheds light on the formation of reptilian ear, a single hole formed of a single bone. But in mammals there is an external growth, pinna, for collecting sound waves. The new fossil, *Mastherium*, about

123 million years old is that of a small animal of terrestrial habitat that gives clues to how mammalian ear evolved. (The Hindu (Daily), 15-10-2009).

Points to ponder as to whether it was Evolution or the handiwork of God:

1. Who created for the first time a hard-protective covering or a hard shell of the egg?
2. Who first created four-legged reptiles on land?
3. Who first created epidermal scales and a dry skin to check loss of water from the body by evaporation?
4. Who first created claws for protection?
5. Who first created the mechanism of fertilization inside the body?
6. Who first created two special neck vertebrae to facilitate head movement?
7. Who first created tubular structured quill in the birds to help them fly in air?
8. Who provided for the first time a bucket like structure filled with sea-like water in the mammals?
9. Who first formed male and female sex organs separately in individuals making them either male or female?
10. Who first created suckling mammary glands?
11. Who first created warm blooded animals?
12. Who first created a flap or epiglottis over tracheae or glottis to prevent food from entering tracheae?
13. Who first created two salivary glands in the mouth?
14. Who first created a diaphragm separating thoracic cavity housing the heart and lungs from the abdominal cavity?
15. Who first created moveable eyelids?
16. Who first formed separate openings outside the body for digestive tract and urino-genital duct?

A variety of Life Forms owe their origin to God, the Creator.

CHAPTER VII

Man, the Final Product

Man is the final formation or final product or he was created in the end as, in the words of God, '*khalqan a'khar*' (another creation). He is the last of creations and as such lies at the top of the line of life formations. Henceforth no new animal phyla will be created. The creation of man, the final product, is revealed by God when He said that He created man from clay and fashioned him in good proportion (38: 71-72; 82: 7). The process of formation of man had started from the day God proclaimed that He was about to create man and created the first living organism.

The conclusive focus of the above Verses lies in the fact that God did create man or the 'final product' under a well thought out plan in three distinct steps in three different time spans. In the first place such an animal whose formation from clay has passed through various stages was created that was different from others in the sense that he walked in an erect posture. It was a crudely formed two-legged creature called man the animal. In the second step body features were sharpened, a balance was introduced in the function of body organs and he was fashioned into a better shape. When every function was perfected and the body was brought to good shape, God as a final makeup breathed into him His Own Spirit or bestowed him some of His Knowledge to make him His true representative on earth as a fully formed complete man.

Man, the final creation belongs to the broad family of mammals that hold important position among the animals occupying the land surface. Biologists have grouped the mammals into three types: omnivores, herbivores and carnivores. Omnivorous mammals started as insect and plant eating primitive placentals and passed into primates, or mammals of foremost rank. Early primates were primitive prosimians or animals before monkeys, and later became full-fledged simians or primates in the form of monkeys and apes. Man lies at the top of the line of omnivorous mammals and is regarded as an 'Able Ape'.

Apart from common characteristic features, certain anatomical and structural modifications were introduced in successive creations of various types of primates. Tree shrews or primitive placental animals are supposed to be at the base of human line of formation. Tree shrews had claws in all of their digits. These clawed digits were the first to be gradually converted into nails in the primates. As a beginning, one toe, the second one, was formed with nails in the primitive primates like lemurs and lorises while all the other maintained claws. Some other primitive primates like modern tarsiers were created with two nail bearing toes, the second and the third ones, while the rest digits were clawed. Later on, however, all the primates like monkeys and apes were formed without claws and possessed flat nails in all the digits.

Tree shrews had small limbs but the same have been enlarged in the primates, though their size is variable among various types. Both forelimbs and hind limbs are equal in lorises, but tarsiers have larger hind limbs. Forelimbs are larger than the hind limbs in all the monkeys and apes. Long forelimbs are of great help to them for living on trees. The hind limbs are easily folded and work as a spring, making jumping from branch to branch easier. Rhesus monkeys however are as comfortable in the movement on land as they are on the trees. Chimpanzees have longer forelimbs than all the apes, but the limbs do not extend beyond knees. They too can easily lead an arboreal as well as terrestrial life.

Primates, unlike other animals, are made with forward looking eyes. The change from eyes on the sides to eyes on the front started in prosimians when lemurs were created with eyes slightly more on the sideways than on the front. All the prosimians to begin with were provided with large eyes but later formations in the line of primates, i.e., monkeys, apes and man possessed smaller eyes. Primitive primates had poor vision but later formations were given binocular vision and could distinguish different colours.

An interesting change from monkeys to apes is seen in the presence of cheek pouches, tail and callosities in monkeys while apes are generally devoid of them. Cheek pouches are present in only rhesus, while other baboon monkeys and apes altogether lack them. Ischial callosities or red pads are a unique feature in rhesus and baboons. Baboons have two small, hard callosities, while other apes are devoid of them. Primitive primates together with rhesus monkeys and baboons were formed with a long or short tail but the same is totally absent in apes.

All the changes took 90 million years in making primates from insectivorous mammals of Cretaceous times up to the formation of apes chimpanzees of late Miocene times. Unfortunately, the fossils of primates are scarce and it is difficult to piece together exact history of formation of various groups of primates. What palaeontologists have been able to gather indicates that primitive primates or prosimians like lemurs and tarsiers came into being in Early Eocene. Early simians or monkeys were created in Late Eocene while baboons were formed in Oligocene Period (32.7-23.5 Mya). The ancestors of apes and man or the early apes like gibbons were created in Early Miocene period (23.5-5.2 Mya). The line of apes then developed further by the creation of orang-utans in mid-Miocene followed by gorillas and then chimpanzees around 5 to 7 million years ago.

Man is formed with many features that were already present in apes. Both have nails on all the fingers. They have round face with forward looking eyes. The facial muscles are made in a way that help in the expression of emotions like anger, happiness, sorrow, etc. Their brain is large and provided with an intricate pattern of convolutions. Like other monkeys, they do not have any pouches in the cheek to ruminate the food. They are also devoid of tail and hard thick pads on the buttock. Both are similar in almost all aspects of their anatomy, physiology and endocrinology.

Humans and apes are very close to each other genetically. About 98.77 per cent genes of man and chimpanzees, a member of ape community, are similar. With so many similarities and only a little genetic difference, it is but natural to concede that the Creator has formed man with a slight modification from apes.

The point that man has been formed from ape is indirectly supported by Quran when it is mentioned that some members of an erring fishing community faced the wrath of God and were changed into apes from which they may have been formed. It is narrated that God under His Commandments to Moses had fixed a day of the week as Sabbath or a day of rest. On this day none was permitted to work, labour, go for hunting or even walk beyond a certain fixed limit. The Sabbath was strictly followed and even fish on this day felt safe and came near the coast popping up their heads as they did not fear any danger to their life. But with passage of time the generations of Moses began to disobey the Covenants of God and together with other Commandments disregarded the practice of Sabbath. God tells about the people of a coastal town where some of the habitants, members of a fishing community went against the Covenant and began fishing on the day of Sabbath. This attracted the wrath of God Who changed them into apes: “We said to them, "Be apes, despised." (7: 166). This fact is again mentioned at another place in Qur’an: “We said to them: Be you apes, despised and hated!” (2: 65).

While describing the punishment meted out to some members of a fishing community, God points out in Quran that there were some other members that resorted to greater evil by breaking Sabbath as well as by worshipping idols and evil spirits in place of God. Such persons attracted greater punishment and were transformed into swine (5: 60). The transformation of men into apes and swine was based on the intensity of disobedience. Those that committed lesser sin were transformed into apes while those indulging in greater sin were transformed into swine.

God has mentioned that the punishment was meted out to erring community as an example to their own people and to succeeding generations (2: 66). People of the time themselves saw it happening and it became an example for others to obey God and not transgress in the Commandments lest they suffer horrible punishment. People of the immediate posterity too were witness to men being transformed into apes and swine and were warned not to disobey God.

This type of punishment appears impractical and unnatural. How suddenly everyone is turned into apes covered with hairs and start walking on all the four limbs. God has power to do it and if He decides a matter He just says 'Be', an it is done'. It means anything ordered by Him is bound to happen. But it may be done instantly, or if the matter is governed by certain rules, it passes through the required process and is done accordingly. The transformation of man into apes did also pass through the natural laws of birth fixed by God for the formation into various body shapes. Henceforth every child born to that condemned community was born an ape. This is proved by the reaction of people as mentioned in the two Verses, 7: 166 and 2: 65. Apes or swine as such are not despicable or untouchable. But when they saw a child born to them was unable to walk erect or behaved like a quadruped ape or swine it became a despised creature and was rejected by them.

The transformation may have been possible if natural rules for genes transferring apes into man were reversed. If it was done, every child born to such person will be an ape. The Creator by way of punishment may have just reversed the function of some specific genes in the members of transgressing community. With the result the new born, instead of being human, were born as quadruped apes or pigs.

The question arises as to why God transformed those sinners into apes and swine and why not into rats, rabbits, dogs or donkeys? Only God knows why it was done. May be, it was done because they are close to the same line of life formations and in a sense are close to each other. Swine and apes belong to the same stock of insectivorous mammals that were created in the Cretaceous Period. Hedgehogs of that stock passing through various changes appeared as swine in early Miocene times. Another line of animal formation that started as tree shrews from the same stock after passing through various forms saw the creation of apes almost during the same period in Miocene.

Both swine and apes therefore, though coming through two different lines of creation, are really distant cousins and share marked similarities in respect of their anatomy, endocrinology and physiology. On the one hand the two cousins share these characteristics with each other, while on the other they share similar properties with humans. Thus humans, apes and swine are close to each other in various aspects. Such similarities between apes and men have already been described earlier.

As far as swine are concerned, their anatomic and physiologic character of various organs and systems functions like that of humans. The heart of the pig is similar to man and its coronary system in majority of humans is also similar. The anatomy of the digestive system is somewhat different but the physiology of the digestive process is identical. The urinary system of swine is similar as the anatomy and function of its kidneys is almost like humans. The skin of swine like that of humans is relatively hairless and like man it is tightly fixed to the subcutaneous tissues. Although the skin is thicker and less vascular than humans, its cutaneous blood supply is similar. All these similarities in the anatomy and physiology between swine and man have compelled men to consider swine as donors of heart, liver, pancreas and kidneys.

Beside similarities in the organ systems between swine, apes and humans, there are many other functions that make them quite different from each other. Swine and apes in particular are wide apart in many ways. The arms and legs of swine are of equal size while apes have longer arms. The former walk fully quadruped, but apes move in a semi-erect posture with a typical shuffling walk on outer edges of feet and on knuckles of hand. Apės have all the five digits with nails on each of them. Swine on the other hand have only 3rd and 4th as principal digits, and 2nd and 5th digits are only vestigial forming dewclaws. The body of apes is covered with long, coarse hairs while swine are relatively hairless. The face of apes is slightly protruding but that of swine is more protruding. Apes can show anger, happiness, etc. with their facial muscles, but swine have no facial expression.

Faced with marked differences, swine stand far away from apes and more so from humans. Apes are close to humans in almost all respects. When God ordered transformation of the erring tribesmen into apes and swine, it is but natural that the transformation to apes, that too with minor differences, was only one step backward in the line of successive formation. Swine standing far behind in the line of formation and armed with many marked differences from man required greater changes in the composition for their transformation. This fact is highlighted in the aforementioned Verse where God said that He punished the erring tribesmen by converting them into apes and swine. Those who were lesser culprits were transformed into apes but if God willing had a chance of early re-transformation into man. while those deep in sin were made swine or put into a distant place in the line of life formation. But those deep in sin and transformed into swine stand far behind the man in the line of creation. In such cases the reversal of genes is required for going many steps backward with little or no chance of re-transformation into man

Man may be similar to apes in many aspects, but there are two major traits that differentiate him from apes. Man has a smooth, almost hairless skin while the body of apes is covered with coarse hairs. The genes of hairless skin were already present in apes but they remained dormant, while the swine coming from the same stock had active genes were thereby created hairless. Man too, while being created from the same stock, needed the activation of such genes to give him a hairless skin.

Man, a Bipedal Animal

The other major structural change distinguishing man from not only apes but from all other quadrupeds is bipedalism. It needs genetic modification in the body make up so that the creature can walk on two legs and bear the weight of its body and various organs. A beginning of bipedalism was already made in apes themselves as some of them started moving in a semi-erect position. But, while doing so, they still use four legs with a typical shuffling motion by walking on the outer edges of their feet and using knuckles of hand.

For perfect erect movement the structure of the hip bones was modified. Pelvic girdle is composed of three bones: the ilium attached to the backbone, and ischium and pubis that provide support to the body. These three bones surround the hip socket of the two legs. Apes have a long ilium that cannot provide support to an erect body (Fig. 6). It was therefore changed to a short but expanded one in man so that it can support viscera of the body that was made upright. Ischium too is so located in man that it can take the weight of the body in a sitting position.

Some other changes were also made in the skeleton to keep human body in an erect position. Apes had a curved vertebral column to suit their movement on all the four legs while living on trees. Man was created with a modified column having two forward and two backward curved vertebrae to balance an upright stance. Another change in the skeleton was made by creating smaller arms and longer legs in man to facilitate bipedal movement on land as compared to longer arms and shorter legs in apes that are needed in their arboreal life.

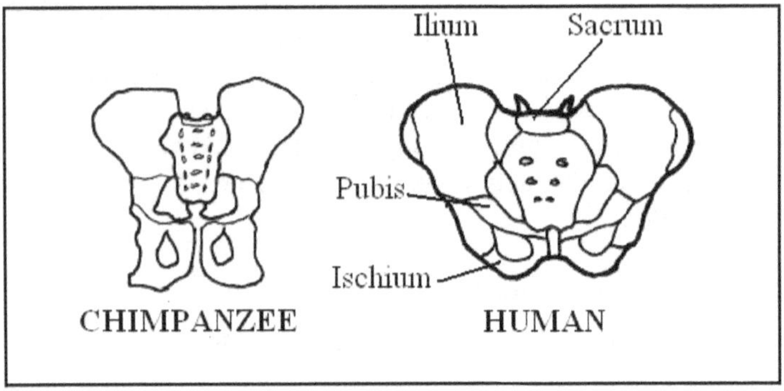

Fig. 6: Pelvic Girdle: Long, narrow in apes and broad, short in humans

In the absence of any evidence by way of fossil records, palaeontologists have so far failed to throw light on any intermediate change of apes into man. Whatever the process of change might have been we just only know that a new creature bearing several features of apes appeared on the scene and began to walk upright. It was a new creation of God known as hominin or man the animal. A chance discovery of some fossilized footprints by Mary Leakey in 1977 in Laetoli, Tanzania is the first non-controversial evidence of the existence of roaming bipedal hominins. The footprints were made by two adults and a child on the volcanic ash turned muddy by rain. The family was perhaps escaping from the nearby volcanic eruption and left their footprints on that muddy stretch. The age of the fossilized layer of volcanic ash bearing the footprints is 3.6 million years old, indicates that hominins were present on earth at that time. Many other fossils of around 4.6 to 1.6 Mya have been found in the east and south of Africa that point to the existence of early hominins there.

The internal and external features of first bipedal animal were refined or shaped in the words of God, "…He refined him in due proportion" until a perfect man, Adam was formed. The refinements may have been achieved by incorporating required modifications in successive hominin species. Palaeontologists so far have found no evidence to know what species were involved in receiving any particular change. It is mainly because the changed features have left no fossils and therefore no evidence in this regard. The fashioning and re-shaping may be morphological, anatomical or physiological in nature. Morphological changes can be done by altering parts of skin, lips, nose, ears or eyes. These are soft tissues and decompose quickly after death leaving no chance of fossilization. Fossilized bones are main source to understand evolutionary changes in vertebrates. However, any minor structural modification cannot be ascertained from pieces of bones that have been found as fossils. Physiological changes of course are not fossilized and can never be found. In the absence of any evidence therefore it is not possible to know what was the nature of the first change and on which species it was achieved. One can only be aware of the nature of change by studying the modifications or improvements leading to the formation of modern man.

Body Refinements

Beside a vertical, bipedal stance and smaller but wider ilia, the endoskeleton of the hominins was further reshaped or remodelled. The toes were shortened and the soles were expanded beyond ankles. The newly formed flattened feet helped the individual in maintaining perfect balance in a standing position. The transition from arboreal life of hominoids to terrestrial life of humans also required some skeletal changes. The hands were no more required for grasping tree branches and legs were not to be used for frequent jumping from branch to branch. The arms and legs therefore were structurally modified to make them useful for new life. The arms were shortened in comparison to legs and the size of digits too was reduced so that after being free from grasping branches they could be used more efficiently for other purposes in handling various objects like edibles or tools, in catching and throwing items of daily use and defence, or in using them in many other ways. The knees were shortened and straightened gradually to cope with a perfect bipedal movement.

Apes had a protruding face that was slightly modified in hominins to give them a better look. The shape was developed in a vertical form (orthognathus) with a vertical forehead. The eyebrow ridges were reduced but the chin was made more prominent. The nostrils too were raised. The jaws were reduced in size and fitted with smaller teeth and less prominent canines. The small gap between incisors and canines or the small diastema that existed in hominoids was removed in hominins. Beside these bony modifications, the face was provided with special muscles that helped hominins possess a better facial expression of their feelings.

The skull was expanded vertically to make it somewhat round as compared to a flattened one in the apes. It also increased in size to accommodate an expanding brain. The cranial capacity of 650 cm^3 in hominoids was raised to 1200-1600 cm^3 in the hominins and Homo sapiens. The brain with greater convolutions gathered more neurons and neural connections to increase intelligence that was put to successful use in procuring food items and defending oneself from various carnivores or from other dangers in otherwise defenceless surroundings of open land.

The senses of hearing and sight were greatly improved. The size of external ear or pinna was reduced and re-shaped in a way that it could catch or receive the surrounding sounds. The whole hearing apparatus was modelled to receive all the fine-tuned, sharp and clear sounds. Likewise, the vision was greatly sharpened and the hominins were able to see three dimensional, sharp and coloured images.

The body of hominins was covered with a smooth skin that had gradually become almost hairless. The hairs however were retained in few parts of the body. On the head hairs were specially formed to grow quickly and continuously. They were also formed to grow quickly and continuously on the face of males as moustache and beard.

Species involved in Re-shaping

The modifications carried out in re-shaping man were not achieved by God suddenly. The process continued for about 4 million years from the time first bipedal animal was created until Adam was born in a final, perfect shape. During this period a large number of species of hominins were created with various modifications or improvements upon one and the other. Palaeontologists or Paleoanthropologists have so far not been able to establish any sequence in the formation of these species as to which preceded whom. The identification however is a difficult task because of the paucity of fossil records. Only a few fossils of hominins have so far been discovered and those too are widely scattered over the world both in space and time. Moreover, in some places the fossils are so intermingled owing to land upheavals that two or more species belonging to different periods of time may be present together at one place at the same time.

The pathways shaping hominins into a perfect man still remain a mystery. A simple way out has been given by Simon Conway Morris in his book, The Crucible of Creation: The Burgess Shale and the Rise of Animals, who suggested to wait for some more fossil finds that may provide evidence for it. On the other hand, Stephen Jay Gould in his book, Wonderful Life, is of the opinion that there was no gradual transformation or a sequential formation, instead it was a jerky process when new animal body or Phyla were suddenly formed. Gould and Eldredge (1977) also point out that in very restricted populations the species may undergo mild changes in morphology in short time. This was perhaps the way as understood by Gould and Eldredge that God introduced 'mild initial morphological changes' at short time intervals during 4 million years.

The jerks or sudden spurts in shaping hominins passing through various species until the creation of modern man are ably discussed by Southwood (2003). Most of the information relating to various species presented here is mostly based on the description given by him. Moreover, beside variations in anatomy, structure and morphology, changing style of living is also taken into consideration. It is possible to group all the known species of hominins into three broad categories: 1) Primitive or early hominins, dependent on forests, 2) Intermediary hominins, partially dependent on forests, and 3) Advanced hominins, independent of forests.

Primitive or Early Hominins

Very little fossil evidence is available for animals created between 5.3 to 4.5 Mya with reduced and broader ilia and walking on two legs. But for later period from 4.6 to 1.6 Mya sufficient number of fossils have been discovered in east and southern Africa that throw some light on the lifestyle of early bipedal animals that collectively are known as australopithecines or 'southern hominins. They were created to move away from forests and begin to walk and live in the grasslands and bushy areas. In the open however there always was danger to life from existing predators. The danger was greater as they did not possess any defensive tools or they were not good runners to save themselves from fast running carnivores. The only defence they had was their long range of sight owing to their increased height. They could see the danger from a distance, could signal or caution other members too about the danger, and could save them by moving away to the caves and/or to the protective cover of the forests. They also therefore did not venture to go far away from the forests. It seems the structure of their arms and legs was not yet changed as it was needed for forest life. In general anatomy and appearance except a vertical stance the primitive hominins remained close to apes.

Australopithecines

So far 16 species of australopithecines have been discovered. These species can be grouped into two types on the basis of the structure and size of the individuals. One group is recognized as 'robust' that has large and strongly built members while the other is 'gracile' composed of smaller and delicate members.

Robust Species

Primates are generally omnivores, but some of them like orang-utans and gorillas as an exception are herbivores preferring fruits in their diet. They were created with strong jaws and huge grinding teeth to crack nuts, uproot tubers and eat even toughest parts of the plant. The plant-eating robust species are placed in the genus Paranthropus. The fossil of one species Paranthropus aethiopicus found in east Africa is 2.5 million years old. Another fossil P. boisei belonging to later period is 2.2 million years old and is also found in east Africa. Two other robust species have been found in South Africa. Unfortunately, no fossils proving their earlier presence have so far been found. Paranthropus continued to live until 1.2 million years ago when they suddenly disappeared from the scene.

Gracile Species

Members of gracile species of australopithecines were widespread over almost every suitable forested patch of Africa. So far 4-2.3-million-year-old fossils of six gracile species have been found here. It is possible to arrange all the gracile species into old or new groups on the basis of their species differentiation.

Old gracile species, Australopithecus afarensis also known as 'Lucy' first appeared 4 million years ago and occupied an extensive area from Ethiopia to Tanzania. for about over a million years. Other gracile species appeared and flourished in various periods of time between 3.5 to 2.3 Mya. They lived in different localities of Africa from Lake Chad in the northwest to Ethiopia in the east or down to Cape Province in the south. Unlike their robust vegetarian cousins, members of gracile species were omnivorous. Their non-vegetarian diet is an indication that they probably lived along the forest edge near the rivers and lakes.

Fossils of two New gracile species, Homo Habilis and Homo rudolfensis (2.8-2.4 Mya) with slightly larger cranium of 552 cm^3 capacity (Paranthopus boisei had a cranium capacity of

513 cm^3) have been discovered in Kenya and Tanzania. They were named Homo or human because the remains were found associated with rudimentary stone tools, a technique indicating human trait. Many paleoanthropologists however disagree in classifying them as Homo merely on account of their tools because chimpanzees too use stones as tools for breaking nuts and for other purposes. The new species were created with increased intellect that was put to greater use in using stones as hammers to break other stones into flakes with sharp edges. Moreover, unlike chimpanzees that use stones only once and then throw them, the new species carried them for as much as 10 kilometres from their original site. The application of intelligence was certainly an improvement, but merely on this basis they could not be placed in the category of Homo. In all other aspects these new formations were still early hominins. They had long arms and short legs with long fingers and toes suitable for climbing and leading a life in the forest. The vertical stance like the other australopithecines was not perfectly erect. The posture was a little stooped and they could walk on the two legs in a jerky way for only short distances. Tim Bromage (1992), an expert on human facial anatomy, with the help of computer simulation of H. rudolfensis skull specimen KNM-ER 1470 has found that its face was not flat like humans, but jutted out like the face of Australopithecus. Other paleo anthropologists Walker and Leakey (1978), on the basis of studies on the same skull no.1470 are of the opinion that this should not be classified as a member of Homo, but should be placed in the genus Australopithecus. True, these new gracile species may not be regarded as Homo, but their slender structure, omnivorous feeding habits, and an increasing intelligence certainly provide a threshold for a major thrust towards this direction.

Intermediary Hominins

A major spurt in the modelling of hominins took place 2 million years ago when God re-constructed the body to be like an erect man, or a true *Homo*. The stooping backbone was converted into two forward and two backward vertebral columns to give it an erect shape.. Soles of the feet were flattened to provide a perfect cushioned grip on land. The legs were enlarged for supporting the body weight and side by side maintaining a proper balance during bipedal locomotion. They also helped in brisk walk or running at ease. Arms were shortened to be more manageable and be useful for multifarious activities. Long fingers were reduced in size to make them free from just grasping the branches and to be put to use in holding, throwing and manipulating various tools and objects. The cranial capacity was raised and the increased weight of the brain was counter-balanced by narrowing the hips so that the balance in the bipedal movements may be maintained.

All the structural changes performed at this stage were necessary to free hominins from the grip of forests and let them lead a life on open land. Their body construction did not permit easy movement on trees but made walking on two legs comfortable. Armed with this facility they began to walk long distances in the open savannah areas in search of food. The presence of hunting axes as far away as 20 kilometres indicates the distances, they may have covered on roaming around. However, as the individuals had not yet developed efficient tools to defend themselves from the predators in the open, they still sought protection in the forests. The dependence on forests gradually decreased in the same proportion as the hunting and defensive tools were developed. Nevertheless, most of the time was spent on walking in the forest in search of fruits, nuts and possible game.

Fossil records have revealed two new species of early *Homo*, H. ergaster and H. erectus. Both had erect posture and differ only in nomenclature. Fossil of *H. ergaster* is a best- preserved specimen of an adolescent boy found near Lake Turkana in Kenya. Beside an erect posture, the new species had thick, projected eyebrows and a slightly protruding face. It belonged to a tall species and calculations confirm that he would have been about 1. 83 metres tall had he reached adulthood.

H. erectus, the other species of erects, had developed special tools that had sharpened edges on both sides of the flakes. The sharp edges and a pointed nose were more effective in killing and cutting the animals. Such Acheulian tools have been found throughout Africa, Europe and West Asia. This fact indicates that these erects were widely spread over these continents from the day they were created until a few thousand years ago.

Advanced Hominins

Africa and its adjacent areas in Asia and Europe witnessed two or three spurts between 800,000 and 100,000 years ago in re-formation of hominins to a perfect shape. The modifications introduced by God gave rise to three new species of hominins that appeared in the scene during this period. These species are arranged as 1) *Homo neanderthalensis* or Neanderthals, 2) Archaic *Homo sapiens*, also known as Cro-Magnons, and 3) *Homo sapiens* or humans.

Neanderthals

Neanderthals were a heavily built, tall species that lived in cool areas of West Asia and Europe around 300,000 to 30,000 years ago. They spread over a large area from North Wales to Uzbekistan in the north and Gibraltar to southern Israel in the south. Proportionate to their robust body, they possessed a slightly large cranium with greater capacity than other species of the time. Unlike humans they had sturdy limbs, thick projecting eye brow-ridges, large nose, and small chin.

These hominins led a better life than their predecessor erectus group. They were no more dependent on forest for protection. There is evidence that they lived in caves that provided better protection from predators during sleep and saved them from the vagaries of cool weather. Their growing intelligence and experience had made them skilled tool makers. Various types of tools were designed from stones of which the sharp-edged, pointed arrow-heads and sharp-edged scrapers speak highly of the skill of the makers.

Neanderthals had arranged themselves into small social groups. The group consisted of members of the family or a few small families may have adjusted to live together. Group living arose as a necessity for better protection and successful hunting. Stone tools were handy in killing small animals. But for killing large mammals like elephants, bisons, horses or rhinos, a collective effort of the whole group was required. The whole group was involved in hunting expeditions. The strategy would have been to firstly stalk the animal and drive it over the cliff or chase them into the bogs and then put the trapped and injured one to an end. The group may have demarcated an area for their activities and may have guarded their territory from encroachment by strangers. The nature of injuries and signs of butchering seen in the fossils indicate that they were perhaps inflicted during fierce fighting between the groups for protection and possession of the territory. It is not known if they ate the dead hominins killed in fighting. But the chances are that they were cannibals and did not hesitate eating any meat available to them. It is not clear whether cannibalism was limited to eating dead hominins of other families or groups or included dead of their own.

It has been suggested that Neanderthals wrapped themselves by animal skins. The idea has cropped up owing to the presence of scraping tools and the special wear and tear patterns often present on the front teeth. It is assumed that the fresh animal skin was prepared for the purpose (of wrapping) by holding it by teeth and then forcefully scraping away the fat and extra meat. The wear and tear pattern on the front teeth is accounted for being the result of their continuous use for this purpose. The assumption seems illogical keeping in mind that only a few persons of the family or a small number of individuals in the group will suffice for the job. For this small use the teeth of the whole population will not wear the way seen in the fossils. Likewise, the presence of scrapers is no proof that they were made only for scraping fat and meat from the skins. The main use of these tools like today may have been for skinning the animal.

One more evidence has been brought forth to prove that Neanderthals did use skin wraps. It is in the form of a bone sewing needle found among the fossils supposedly used for joining the skins to make them wearable. The needle is very smooth, sharp pointed and carries a hole. The finish of the product defies any thought that it could be prepared by tools available to Neanderthals. Moreover, the needle was made 26,000 years ago while the men supposed to have made it were already extinct by 30,000 years ago. Is the needle finely crafted from a 26,000-year-old fossilized bone and recovered with Neanderthal fossils of 35,000-30,000 years back? Is it an ingenious way to prove that Neanderthals used to sew skins for wearing? Was it another attempt like Piltdown forgery? In the absence of any strong evidence therefore it is better to assume that Neanderthals like other hominins roamed about naked. These people once occupying a large territory called Neander land, encompassing southern Europe and south-eastern Asia, gradually became extinct.

Cro-Magnons

Around a million years ago another spurt in re-shaping hominins resulted in the formation of a new species with slightly modified features that somewhat resembled men of today. They had a dome-shaped cranium with a broad forehead and slightly protruding eyebrows. Their cranium capacity of 1600 cc was slightly larger than other hominins of the day. As the new species was not truly human, it was designated as archaic human or archaic *Homo sapiens*, often termed Cro-Magnons.

The fossils of archaic *Homo sapiens* found in the caves of Quafez and Skhul in Israel are 100,000 years old and are similar to the fossils found at Broken Hills. It is thus assumed that this species was created in Africa from where it moved northward into Europe. Almost at the same time when the new species was created, Neanderthals also had occupied Israel as indicated by their fossils found in Tabun. Neanderthals had then moved on to occupy major parts of Europe and West Asia. They had returned (or some continued to live there) to

Israel as seen by their fossils of 50,000 years ago in Amud and Kabara. Thus for 40 to 50 thousand years Neanderthals and Cro-Magnons appear to have lived close to each other in the eastern coastland of Mediterranean Sea. No doubt, Cro-Magnons living in the same cultural milieu, may have shared the same skills and standard of living as possessed by Neanderthals. Both of these species intermixing with each other are collectively known as Archaic *Homo sapiens*.

Humans

Remodelling of hominins by God into a good shape continued until the shape of modern humans was achieved. In the last but one such attempt a new species carrying human features was created that was called man or *Homo sapiens*. The cranium, though a little reduced than archaic Homo sapiens, was made a little higher and curved. The face was formed with a vertical forehead and the brow ridges were reduced specially in the middle. The appearance was made further attractive by adding a more prominent chin. Fossil evidence indicates that the species carrying perfect human features was created around 100,000 years ago.

These *Homo sapiens* were better artisans. They made a variety of well-designed and better chiselled axes, scrapers and borers. Beside stones, they had also started using bones and antlers in making longer, narrower and sharper tools. A great leap in tool making was achieved when a handle was attached to some tools. It not only increased the force of impact and the reach of attack, but also saved them from the danger of close, hand to hand fight. The discarded tools and some other attractive items were occasionally used as ornaments. It indicates that they did possess some aesthetic sense.

True, early humans or early *Homo sapiens* were skilled tool makers, good hunters and fighters, and possessed a little aesthetic sense. Their style of living however in many ways was not different from that of animals. There is no evidence to show that they did not live naked or that they did not eat the dead of their own species. The guess is that early humans did lead a life like animals. They lived naked, practised cannibalism and indulged in uninhibited act of procreation. Early humans that way, in fact were animal *Homo sapiens*.

Points to ponder as to whether it was Evolution or the hand/work of God:

1. Who created for the first-time long hands and feet in the primates?
2. Who first created five fingers?
3. Who first created opposite toes that help in grasping and grabbing?
4. Who first created highly convoluted brain in the primates?
5. Who first created only two teats in the primates?
6. Who first created forward looking eyes?
7. Who first formed a binocular vision?
8. Who re-structured pelvic girdle and made it perfect for supporting an erect body?
9. Who modified vertebral column with two forward and two backward curved vertebrae to balance an upright posture?
10. Who formed flattened feet with soles extending beyond ankles to have perfect grip on land, to maintain perfect balance in standing position, and to make it easy for brisk walk or for running?
11. Who shortened knees and gradually strengthened them to cope with perfect bipedal movement?
12. On shedding arboreal life and starting life in open land, who brought about useful structural changes in the skeleton of hominins by shortening arms in comparison to legs and by reducing the size of digits in fingers?
13. Who provided special muscles in the face for better facial expression of feelings of joy, sorrow or anger?
14. Who expanded skull vertically to give it a round shape and increased its capacity to accommodate an expanding brain?

15. Who made a smooth skin that gradually became hairless except in a few places of the body?

16. Who made hairs on the head to grow quickly and continuously?

17. Who made males to possess beard while females were spared from it?

18. Who re-formed Neanderthals with sturdy limbs, thick projecting brow-ridges, large nose and small chin?

19. Who re-shaped Cro-Magnons with a dome-shaped cranium, broad forehead and slightly protruding eyebrows?

20. Who re-modeled hominins to look like modern man by making cranium a little higher and curved, forming a vertical forehead and reducing brow ridges specially in the middle?

21. Who made face further attractive by adding a more prominent chin?

God has created and fashioned human body in good proportions and better shape.

CHAPTER VIII

God Conscious Perfect Man

During the first 1.8 million years Before Present (BP) of the Early Pleistocene (2.58–0.8 Mya) Africa witnessed dry conditions leading to loss of vegetation and means of sustenance. Local hominin population began to move out in search of greener pastures. At the same time anatomical reformations as required by God were taking place in each new species involved in migration. On the basis of fossil records found at the place of destination, five such movements over different periods of time can be recognised. Three such radiations out of East Africa went towards West Asia and Southern Europe, while two were made towards adjoining areas into Arabia.

The Levant or the eastern coastal area of the Mediterranean Sea comprising of Turkey coast, Syria, Lebanon and Israel was the main centre of attraction. Here the Dead Sea Rift Valley created around 2 to 3 million years ago has been a biogeographically suitable place for living and for providing easy route to various streams of hominins moving out of Africa. Fossils of different species found in this region point that most of them had settled here and took advantage of natural resources of the area. The first group radiating out of

East Africa reached this Valley about 1.6 Mya. Other streams of migrants came around 780,000 years ago. Fossil remains discovered in caves of the Jordan River Valley belong to different hominin species of between 80,000 and 130,000 years ago. Beside the remains of East African hominin species, there also exist many fossils some 70,000 years old of Neanderthals. These were residents of Central and Northern Europe, but faced with increasing cold of the Great Ice Age, were apparently forced to move out and settle in this valley.

Migration to Arabia

Some groups of hominin population that moved out of East Africa towards Levant region, turned to go to Arabian Peninsula. It was the time when the whole world was under the grip of Great Ice Age. The glaciation had resulted in holding up large quantities of water in the form of ice sheets in north Europe and north Asia. The Ice Age characterised by its alternating cold and warm spells had great impact on natural vegetation everywhere in the world. In the interglacial period with favourable warm and wet climate, the forests became luxuriant and spread continuously over a large area. During glacial period with cold and dry climate, the same were reduced into isolated patches of vegetation known as '*refugia*.'

Arabian Peninsula during interglacial days was much wetter and its greener pastures and many lakes and rivers presented great attraction for people of East Africa. The attraction intensified during fourth glacial period when the level of Red Sea further decreased. The reduction in the water level all over the seas in northern hemisphere is estimated to be 120 metres that of today. During this glacial period the Red Sea had become shallow and contained only a small quantity of water. It provided an opportunity for *Homo sapiens* to cross over from Africa and settle in parts of Arabia. Migration to Arabia is supposed to have taken place in the last glacial period starting about 127,000 years ago during the Upper Pleistocene times.

The first batch is supposed to have crossed shallow and narrow portion of the Red Sea at Bab al-Mandab around 125,000 years ago and proceeded towards UAE. Some members of the same group, however after crossing Bab al-Mandab took a northern route along the narrow western coastal plain and occupied a wider plain near Jeddah at the first break of the high western mountain. The migrating *Homo sapiens* on arrival may have occupied this favourable 'refugia' about 80-70 thousand years ago in some isolated patches of natural vegetation in the low-lying parts.

The two centres occupied by African migrants, viz. Levantine Corridor and Arabian Peninsula were ready for the formation of new and final species of hominins that may be anatomically complete. Levant was too crowded with constantly moving populations and therefore did not provide isolation necessary for the formation of new species. Europe on the other hand was alternatively cool or cold inhibiting variety in environment that helped speciation.

Diversity of natural environment and an isolated existence available in Arabia are favourable factors for speciation. Homo sapiens occupying one such suitable 'refugia' for 40 to 50 thousand years may have witnessed speciation when Adam was formed as a new species.

Final Anatomical Correction

Language

All animals have power to express anger, happiness, threat or submission, but they lack in power of oral communication. Early Homo sapiens, though apparently similar morphologically to modern humans, could talk with each other in a limited way in the form of some specific noises. It was because of a persistent shortcoming in the formation of their cranium.

A minute examination of their skull reveals that like Neanderthals they had a reduced area for accommodating that portion of the brain which controls speech. With a small part of the cranium available for this purpose, the brain remained restricted and less developed. Because of it early Homo sapiens could not have expressed their feelings coherently.

Besides cranium, larynx in early Homo sapiens remained stuck high on the throat and thus was not available for modification and modulation of sound. In modern humans the larynx slowly descends to its proper place in a few months after birth and the child begins to frame words and starts talking. Oral communication is possible when the areas of brain connected with the formulation and delivery of speech are well developed and the lips, tongue and the larynx are perfectly aligned to form words.

Early men or Homo sapiens continued to live with this shortcoming for about 50 to 60 thousand years until the final and finishing touches were made to remove the defect. It was around 50,000-45,000 years ago when God selected a suitable foetus and introduced into him the required modifications. Its anatomical deficiency was genetically removed by enlarging a part of the cranium so that it could accommodate an increased volume of speech related brain.

Side by side the larynx was made to move posteriorly to occupy its proper place at the required time and help the first member of the new species to talk smoothly. Adam, the first member of the new species, was born with this last anatomical correction and God taught him speech (55: 4). Soon after birth he started to formulate words, join them and began to talk.

A similar contingency arose when God set the larynx at place before the birth and made Christ to talk soon after birth and expel people's doubts about the piety of Mary: "… He has given me the Scripture and has appointed me a Prophet, and has made me blessed wheresoever I may be…"(19: 30-31); "Peace on me the day I was born, and the day I die, and the day I shall be raised…" (19: 33).

Non-Genetic Perfection

Spirit of God or Rooh: A Gift

God created the Universe and all it contains under certain predetermined rules known as Natural Laws. Each item has been formed accordingly with in-built information of these natural laws and functions. Living things too were provided with this information from the day first single living cell was created. Each cell carries a mine of information in the form of DNA about the rules and parameters fixed for its function. The in-built information contained in the DNA is released automatically for the purpose it is required. Based on the specific information for specific body part, the enzymes start production of those proteins that are required for their formation. Even if the body part is damaged the DNA helps reproduce proteins required for replacement or rejuvenation of the part, provided the natural laws permit. For example, the damaged skin is easily re-formed, or dead blood cells are quickly replaced by new ones. The coded information contained in the DNA of living cells works mechanically like an automatic machine in the formation, development and function of various parts. All living things including man are governed by these genetic natural laws throughout their life.

When the formation of man passing through various physiological, anatomical and morphological corrections in its built and shape reached its final stage, God performed a great and final act of benevolence to mankind by breathing into Adam a part of His Spirit of consciousness, knowledge, and understanding of various things (32:9; 2:31).

The infusion of Spirit was the final gift of God that distinguished man as superior to Angels. Man possessing specific faculties, aptitudes and capabilities became a dominant species and God's true Vicegerent on the earth. The superior position demanded recognition, and God commanded His Angels to pay their respects to Adam: "When I have fashioned him (in due proportion and shape) and breathed into him My Spirit, fall you down in obeisance for him..." (15: 29).

In order to prove the superiority of Adam, God placed all the things before Angels and asked them to describe what they are but they expressed their ignorance. God then asked Adam to inform them the nature of things and it was done (2:32-33). Adam became the first person to have talked soon after birth.

The similarities between Adam and Jesus talking soon after birth are indirectly implied when God mentions the similarities in the birth of Adam and Christ. It is mentioned that Adam was created from dust when God "said to him: "Be" and he was." (3: 59). Mary, questioned about how a child will be born when nobody has touched her, God said when a thing is decreed, it is bound to happen (3: 47).

The conscious, knowledgeable and thinking mind was not the outcome of any genetic modification, but in fact was the result of non-material, non-genetic spirit of God that, when introduced, made the brain a powerful, thinking, and guessing store house. The non-genetic Spirit of God was introduced apparently into Adam's mind, but it was to continue into all the descendants of Adam.

Besides information and knowledge, God consciousness and thoughtfulness was also entered by God in the brain. It gave humans the ability and awareness to discern right and wrong, filled it with power of reasoning to form conclusions, and provided the capability to understand and express things clearly.

Man was created to test how far he remains just and truthful in activities. The faculties of awareness and reasoning were provided to come to a just decision about good or bad. It was the way to know whether he be grateful or disbelieving (76: 2-3).

The faculties infused into Adam had come with their own important characteristics. They were not genetic in nature but by the Will of God continued to pass on into the descendants of the new species. The store of information and knowledge passed on to each child did not function automatically like the DNA does and was not available easily to each and every one. It was left to the discretion of the individual when he grows up to get as much information as he could extract. For it he was provided with certain faculties like hearing, feeling and power of understanding to gather information and derive conclusions (32: 9; 23: 78). Ears, eyes, nose and other sensory organs of the body are natural tools in hand to collect data. The knowledge of things and the information about their functions is available to everyone at his own free will and effort. Anyone interested in knowing the nature of things can activate the consciousness in the brain, analyse the collected data and retrieve the required information according to his/her will. The variety in the information obtained is thus the outcome of differing interests of the user.

The retrieved information and knowledge of various items is imprinted in the memory for future use or for sharing with other individuals. The knowledge and experience thus gained is transmitted from generation to generation through signs or by oral communication, and by observation of elders' activities. The store of retrieved information increases with each passing generation and is used to produce philosophers, scientists, politicians, artists, etc. Depending on the effort and interest, and armed with cumulative retrieved knowledge, any person can delve deeper into the store of knowledge endowed by God, take out what he wants, and excel in his field. The store of His Spirit is always available to the descendants of Adam as an instinct to judge good or bad.

With the two modifications accomplished, i.e. making him ready to talk and infusing in the mind consciousness and knowledge, God created a God-conscious, Anatomically Perfect Human (APH) or Modern Human, Homo sapiens sapiens. The new product, Adam was the ultimate creation and as such was the culmination of all biological changes with regard to humankind.

Early Homo sapiens were leading an animal life, but after many improvements, the creation of Modern Man completely altered animal lifestyle.

Cultural Attainments of Perfect Humans

Food Habits

Early races among the descendants of Adam were hunters and food-gatherers. Besides fruits, seeds, shoots and roots, meat was their main diet and they were efficient hunters. They had crafted better tools for hunting made from flint, bone and wood. The use of wooden handles in axes, hammers and spears had increased their killing power besides keeping them safely away from close contact with the hunted animal. With the help of somewhat sophisticated hunting tools they could easily kill even large animals.

The preceding race of Neanderthals were heavy meat eaters. It is not known whether they used animal meat or were also cannibals obtaining it through group fights or local skirmishes. Quran has revealed a situation when God declared that He is about to create man who will be His Vicegerent on the earth. The angels asked will He place therein one who will make mischief therein and shed blood while they always praise and glorify you. God replied that they do not know what is known to Him (2: 30).

The angels were witness to what was happening on the earth and out of curiosity dared to inquire if these mischievous people (hominins) that kill each other were fit to be honoured as God's Vicegerent? Fighting and killing each other as such is not so repulsive as to encourage angels to express their anxiety to God. It was a routine matter to establish the supremacy of different tribes over their territory in pre-Adam races of hominins. They had to fight and kill others to preserve their right. There may however be some hidden interests in generating the fight. They were 'mischievous' people and could have by scheming intentionally created circumstances to start war. The temptation behind the fight may be to easily obtain prized, tender and tasty meat of the killed ones. The courage of the angels to query God may be the outcome of this abominable act of fighting and eating the meat of members of their own species. God consoled the angels that only He knows the future and that they do not see how the things will come out afterwards. It can thus be inferred that the mischievous, scheming hominins residing at the time Adam was born, like their preceding hominins, were cannibals. The practice was a common phenomenon. It cannot however be ascertained if they ate every dead individual, or they spared members of their own tribe and cannibalized only others.

Beside meat obtained from hunting or fighting, later *Homo sapiens* like their predecessors also depended on forest produce for their survival. The process of food-gathering became more efficient when new species of *Homo sapiens sapiens* possessing conscious mind came into being. The food items then were collected and kept in stores for use in future. Some thousands of years later around 10,000 years ago the plants were domesticated and cultivation started to provide continuous food supply.

Body Wraps

Homo sapiens were owners of an increasing intelligence that coupled with cumulative experience in tool making helped them lead a practically easy life. They lived in caves and used fire to ward off animals and protect them from cold. They used finely crafted tools to save them from attack by others and to kill dangerous and other animals. Their intelligence however was still incomplete as they were devoid of awareness about their achievements and led a life like unconscious animals. This shortcoming was removed in the new species of *Homo sapiens sapiens* or in the perfect man that, beside the intelligence gathered by animal *Homo sapiens*, was created with an added gift of knowledge, consciousness and understanding. This gift is available to everyone but a person had to make efforts and strive hard according to his capacity to partake this knowledge. In the beginning however, as mentioned earlier the rule laid down by God was relaxed for the first member of the new species when Adam was born with full knowledge of existing items. For knowing the nature of many others, he was governed onwards by the general rule and had to wish and make efforts to seek knowledge.

The first thing Adam learnt himself was the awareness about sex, though it was acquired in a bad and hard way. Adam was living an animal like life with his parents and other members of the tribe unaware of the fact that he was a different species and was made superior to others. All were placed in a garden where every item of need was available in abundance and they were passing their life at ease. Adam and a female member, Eve, had come close to each other and were living like life partners. Both of them were ordered to dwell in the Garden; and eat of the bountiful things therein as they wish but should not eat the fruits of a specific tree (2: 35; 7: 19).

Adam and Eve continued to live in the Garden of bliss and enjoyed all the facilities available there. But, like animals, they were still unaware of sex and procreation. Coaxed by Satan and out of inherent curiosity to know more, they went against the order of God and tasted the fruits of the forbidden tree. No sooner had they tasted it, the knowledge of sex got suddenly revealed to them and they became conscious of their nakedness and were filled with shame. The knowledge of sex and awareness of nakedness and shame was the first item that Adam extracted by his own efforts and by his own free will from the great store of knowledge. As the shame became manifest to them, they began to sew together the leaves of the Garden over their bodies (7: 22; 20:121).

The knowledge of shame nevertheless was of immense advantage not only to Adam but it paved the way for his descendants too for getting a cover to hide the naked body. All the hominins including more intelligent *Homo sapiens* were not aware of their nakedness and therefore never used any extraneous material to cover their bodies. It was only when Adam and Eve became conscious of their nudity that they immediately tried to cover naked body by leaves, the only material available to them for this purpose. The practice gained ground and people started using animal skins followed by woven clothes for hiding their shame, and for adornment of their personality (7: 26).

Burial of the Dead

Some fossils of Neanderthals found in the form of complete skeleton indicate the practice of burial of the dead individuals. It is suggested that these people not only buried their dead but they would have protected the body for some time from wild scavengers. The presence of some stone tools or animal and plant remains near the skeleton is taken as another proof of burial because they could have only deliberately been placed. Some Scholars further venture that the position of the body placed on its sides with legs folded up with tools and food items lying close by is good evidence to prove that Neanderthals were religious.

These presumptions however are contradicted by the information given by God in the Holy Qur'an. A verse in the Qur'an informs that the practice of burying the dead started only after the new conscious species was created, the knowledge of burial was acquired by Homo sapiens-sapiens when one son of Adam murdered his brother. God then sent a raven that scratched the ground to show him how to hide his dead brother's body (5: 31).

All the hominins before this event whether *ArchaicHomo sapiens or Homo sapiens,* in the beginning were ignorant of the fact and did not bury the body. Perhaps the close members of the dead person (found as fossil) left the body and other belongings in the cave and went away after sealing the cave mouth by pebbles and stones collected from nearby. It may be taken as a crude form of burial but in fact it was an unconscious effort devoid of digging the ground and then interning the body. It may thus not be regarded as a proper burial. It may also have happened that the cave suddenly collapsed burying the owner sleeping with folded legs together with his belongings of daily use. If the death had occurred outside the cave in fighting or otherwise, the body might have been cannibalized soon or eaten by hyenas, vultures or scavengers and the bones may have scattered in the area. Concern for the dead and the practice of burying the body started only after the *Homo sapiens sapiens*gained knowledge and awareness of the bond of love between the members of the family and of the pain on account of the loss of the dead loved ones as mentioned in the above Verse.

Time of Birth

Modern humans were formed after passing through a series of biological changes from the day first bipedal animal was created. The changes made over time resulted in the formation of many sub-species of hominins. Most species lived together for hundreds and thousands of years and therefore it is difficult to find exact time of their creation. Archaeologists, basing their judgement on some major specific changes in the fossils, have suggested broad range of a period when new species were formed. For example, it is known when bipedalism started, when perfect erect posture was obtained or when a 5 to 6 feet tall being came into existence. It is also known when protruding facial structure changed into flat and long but narrow face or when a thick, low forehead with brow-ridges gave way to a vertical forehead with low brow-ridges. Fossil records show in general the period when robust species of hominins occupied the earth or when tender ones walked on the surface. The broad age of the presence of Australopithecines, archaic *Homo sapiens* (*Homo habilis and Homo neanderthalensis*), and *Homo sapiens* like Cro-Magnons is also known.

The determination of exact period of formation from fossil records gets difficult for the last species in the series because it was created with only a small, though most important, modification. The minor change remains indistinguishable in the fossils with the result the possible time of formation continues to be a point of discussion. The problem is further complicated by the fact that many species of later hominins like archaic Homo sapiens, Homo sapiens, and the latest Homo sapiens-sapiens continued to live together for a long period of time. A separate date cannot be fixed for the creation of new species from an amalgamation of almost identical fossils of different species present at the same period of time. It has been suggested that the new species of modern humans appeared on the scene about half a million years after the formation of Homo sapiens.

Apart from fossils, modern techniques of molecular biology developed recently in the last decade of the 20th century have been successfully utilized to find the true age of any biological change leading to creation of new species. Two approaches are adopted for this purpose. Studies of mitochondrial (non-nucleus) DNA, which is inherited through females, reveal details of maternal lineage. The other Y-chromosome (nucleus) DNA, inherited through males provides information of paternal lineage. In the light of these differences, geneticists have collected both non-nucleus and nucleus DNA from a wide range of modern humans occupying a wide variety of habitats. The genome for mitochondrial DNA reveals that modern humans are descendants of a common female ancestor that lived about 200,000 years ago. Studies relating to Y-chromosome DNA that reveal paternal lineage however point to a more recent origin of humans.

Precise time of the birth of Adam can be obtained if microbiologists prepare a genome of that gene which resulted in language related anatomical modifications and started a new talking species.

Taylor (2008) points out that the delivery of words or speech is controlled by areas of brain located in the left side of the cerebrum: i) The speech motor cortical area controlling the delivery of speech, and ii) the area storing auditory, visual and verbal information.

These areas were involved in the final modification carried out by God in the creation of a perfect man.

The part of skull holding these portions of brain was enlarged to accommodate increased size of language related brain so that the new species could talk coherently. If genes responsible for the formation of perfect skull and speech related brain in modern humans are identified and the genome of the same prepared, microbiologists can certainly arrive at a precise date when the last biological change occurred and Adam, the first perfect man, was born.

Child Adam and his clan were faced with increasing aridity and gradually decreasing means of sustenance. In order to ward off this misery God commanded Adam to dwell in His Garden and eat of the bountiful things as they wished." (2:35; 7: 19). It was a perfect 'refugia' of lush green forest full of a variety of animals where none will remain naked, hungry, or thirsty and be exposed to sun's rays (20: 118-119). As the place may have been located at a higher level, the weather was cool and pleasant. This place was occupied not only by Adam and his wife, but the whole community settled there.

Location of this place at a higher level of any mountain range is evident from a Verse in Qur'an that on defying the directive of God, Adam and everyone were ordered to go down from the 'refugia' or the 'Garden' (20:123). This Verse also clearly points to the fact that the place was occupied not only by Adam and his wife, but the whole community was present there and all were asked to go down to the plains.

Place of Birth

It is not possible to pinpoint exact place where Adam was born. Qur'an reveals for certain that Adam was asked to pass early growing period in an area rich in flora and fauna full of all the means of sustenance. But in the absence of any fossil finds in Arabia there is no way of knowing where the event took place. If, however, some indirect information available in Qur'an, in some local traditions, and in scientific investigations are considered, one can arrive at certain conclusions in this regard. The indications for the purpose are:

i) Arabia is privileged to possess first house ever built on the earth and according to tradition is supposed to have been built by Adam on the place and within the precincts indicated by God. It is a cube-shaped construction known as Ka'ba located in Mecca (originally named Becca or Bakka) at a distance of 75 kilometres from the Red Sea coast. God has proclaimed it to be His Own House at Bakka, full of Blessings and guidance for all kinds of beings (3: 96). It is befitting that most prized of all the creations, and in fact the representative of God on Earth, Adam should have been created in the environs and in the land of His Holy House.

ii) There are two other places in Arabia called Arafat (meaning knowledge) and Mash'ar Al Haram (meaning consciousness and understanding) that signify the two qualities with which Adam was created as a perfect man. Arafat at present is a dry sandy plain situated at a distance of 15 kilometres from Mecca, and Mash'ar Al Haram is a Mosque that lies close to it. The two places by their own names represent the region where God bestowed on Adam His Spirit and created him here as a perfect man full of Arafa (knowledge) and Mash'ar (consciousness).

After expulsion from the Garden and with no sense of direction, Adam wandered for long time over unknown places in search of his place of birth. He was continuously in grief for his act of disobedience and constantly praying for mercy. After wandering for many years, one fine morning he found himself in Arafat. Here he stood repenting for his sins and ultimately God not only pardoned Adam, but he was granted Prophethood (2: 37). The plains of Arafat and Mash'ar Al Haram were sanctified by God as places where knowledge, inspiration, consciousness and mercy were bestowed on Adam.

iii) There is a chain of rocky hills called Jabl al-Rehmah near Arafat plains. The hills may have formed a continuous range of mountains with higher elevations 50,000 years ago. During those days they might have been a perfect 'refugia' of lush green forests full of a variety of animal life compared to the shrinking vegetation and diminishing means of sustenance in the plains. God the Merciful permitted Adam and Eve together with their clan to dwell in His Garden. The Garden since then began to be known as the Mountain of Mercy or Jebel Al-Rehmah because it was here that God bestowed His Mercy and Benevolence on Adam.

iv) When Adam was misled by Satan God expelled him and all others from the place of 'Rehma' and forced them to go down (2: 36). The expulsion was sudden and may have been caused by any natural calamity in the form of volcanic eruption or forest fire. As one approaches this mountain from the north, the constitution of rocks changes from sandstone to trappean formations. The north western interior of the Peninsula is a sandstone plateau of about 1200 metres elevation. South of this plateau there are lava fields that topping over mountains south of Medina overflow almost up to the western coast. It is just possible that at some point on the top of mountain where Adam and his tribe were enjoying the facilities of the Garden, suddenly the volcanic eruption started spewing lava, ash and fire. Faced with the severity of the calamity everyone ran away as fast and as far and anywhere as he can go to save oneself.

v) All the places associated with major events in the life of Adam have been sanctified by God and have been made essential part of annual Hajj pilgrimage. Performance of Haj has been made compulsory for every Muslim who can afford, and in order to fulfil its requirements the pilgrim must visit, beside other places, those that commemorate the presence of Adam:

a) Standing (woqoof) in obeisance to God at Arafat is an essential part. Prophet Muhammad mentioned that Al-Haj Al-Arafat, meaning Haj is the name of Arafat (*Al Mustadrak al Saheheen*, 2/305 Number 3100). God will forgive every sin and will shower His Mercy on all those that ask for it standing in Arafat like Adam once did and was pardoned for his act of disobedience. Any one not standing in obeisance to God in Arafat will forfeit his right of Haj performance.

b) After leaving Arafat you should celebrate with praises of Allah at Al-Mash'ar Al-Haram (Sacred Monument), and celebrate His praises as He had directed (2: 198).

c) "Then let them complete the rites prescribed for them, perform their vows, and perform circumambulation of the Ancient House (Ka'ba)." (22: 29).

vi) The final resting place of Adam is in Mecca while Eve is buried in Jeddah.

vii) Since there is a very small difference in the anatomy of Cro-Magnon and modern man, it is difficult to pinpoint if the fossils of that age belong to either of the two. Recent genetic studies have however, come to the rescue and shown the age and place where the first ancestor of modern man was born. Studies relating to mitochondrial DNA (mt DNA) tracing the maternal lineage led to a female ancestor that lived in Africa about 160,000 years ago (Cann R. L. et al, 1987; Vigilant L. et al., 1991). As against it, Y-chromosome DNA leads to paternal lineage. A research based on Y-chromosome testing of 10,000 people from indigenous modern population from Europe, Near East, Arabia, and all over the world conducted by the Geno graphic Project points to a common male ancestor born within Arabia from which all Y-chromosomes have descended (Magazine *Natuurwetenschap en* techniek, 2009). In fact, all the migratory routes prepared for this purpose radiate from Arabian Peninsula indicating it to be the birth place of Adam.

All the supporting evidences and genetic studies point to the fact that the environs of Mecca in Saudi Arabia were the place where first Anatomically Perfect Human (APH), Adam was born and also the place where he passed major part of his life.

Points to ponder as to whether it was Evolution or the handiwork of God:

1. Who enlarged a part of the cranium to accommodate an increased volume

2. Who made larynx moveable to occupy its proper place at the required time?

3. Who bestowed consciousness, knowledge, understanding and thinking mind to human beings?

4. Was conscious mind, full of knowledge and thinking, the result of genetic modification or God-given?

5. Who first provided consciousness about sex and nakedness to cover-up?

6. Who first guided man about burying the dead?

The power of thought, understanding and knowledge are a Gift from All Knowing God, the Perfectionist

CHAPTER IX

Dispersal of Perfect Humans

Arabian Peninsula at the time Adam was born was much wetter than it is today. There was a large network of lakes, rivers, and valleys. Greater part of the mainland was covered with green grasses and there were ample means of sustenance. With the passage of time weather gradually started to get warmer and increasing aridity affected natural resources. Side by side, the population of newly created Perfect Modern Man or *Homo sapiens sapiens* also increased beyond the carrying capacity of the land. The overflow of the population forced them to move out of the mainland in search of better opportunities of survival (30: 20).

In the absence of any direct evidence to ascertain the route of dispersal of newly formed humans, scientists have tried to solve the problem through genealogical DNA testing of varied population groups around the world. Since Y-chromosome is present unchanged in the nucleus inherited by each of the cells of all human males, its genealogical DNA test (Y-DNA Test) determines patrilineal ancestry. On the basis of these studies, some common genetic markers of Y-chromosome were finalised for samples collected from various locations all over the world. Markers reveal the paternity or lineage of men and where and when they first appeared.

The genetic markers indicate that early ancestors of modern humans crossed the Red Sea and spread over the western coast of the Arabian Peninsula 60,000 years ago. They further spread on the interior plains where a new species of Anatomically Perfect Humans was born in the form of Adam sometime between 50 to 45 thousand years ago. The genetic markers of Y-chromosome point to the presence of descendants of Adam over a large part of the mainland around 45,000 years ago. A useful study based on markers of many living people showing the dispersal routes of descendants of Adam, the first Perfect Man, is given in the Magazine, *Natuurwetenschap en techniek*, October 2009 (Fig. 7).

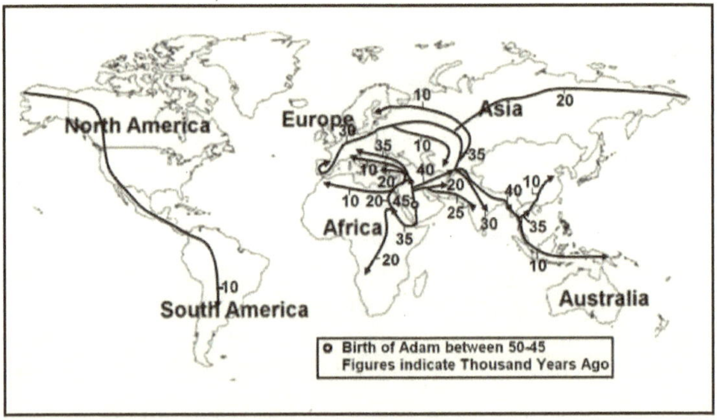

Fig. 7: Dispersal of Descendants of First Anatomically Perfect Human: modified after "Tracking Y Chromosomes Through Time", from Magazine, Natuurwetenschap en techniek, Oktober 2009, cf Random notes: geographer-at-large blog: http://geographer- at large. in/2012/02/ tracking-y-chromosomes-through-time. html.

Any migration of population from the Arabian Peninsula towards the south or east was restricted by the presence of the Arabian Sea and the Persian Gulf. The easiest way open was the land route towards north where extensive plains in the 'Fertile Crescent' were waiting to absorb them.

The bulk of migrants adopted that route and went northward to occupy West Asia. From there some of them moved further westward to spread over lands in Europe while some moved eastward to Central, Eastern, and Southern Asia. The movement to the northernmost parts of Europe and Asia may have been restricted by the extreme cold of the time as these parts were under the influence of the Great Ice Age. These lands may have been occupied only after the climatic conditions got better and the area became hospitable for living. Modern men were then able to cross Bering Strait and spread over North and South America.

Dispersal in Asia

Migrants moving northward from Arabian Peninsula about 45 thousand years ago reached the Fertile Crescent. These fertile lands spread over an extensive area in present day Iraq, Syria, Israel and Jordan. Beside the attraction of these fertile lands, the valley of a major river, Phorat, was available to provide easy access to Fertile Crescent. The River Phorat rising from a point in high Sirat mountain range in Hejaz region and winding through the environs of Mecca flowed northward and drained into a large lake. This major river of ancient times is now lost but its dry bed can be seen in satellite imageries of the area. The population moving through this valley reached and settled in the plains of Iraq.

In due course of time the increase of population forced few groups to move westward of which some settled in the western part of Fertile Crescent or the Levant region while some continued their journey up to Europe.

The Levant Region was occupied by newly arrived modern humans around 40 thousand years ago. There are shreds of evidence to show that some species of Archaic *H. sapiens* at that time were already present in the Levant region. Of these, Neanderthals in fact were spread over the whole of habitable parts of Europe.

Archaic *Homo sapiens* present in the region were similar to the newcomers in built and shape but in comparison to them the newcomers, *Homo sapiens sapiens* or early modern humans were delicate and weak. As the newcomers were alien, the locals resisted any co-existence with them. But in spite of their physical weakness, the newcomers were superior in knowledge and well-versed in warfare. They were also better tool makers, planners, and strategists and threatened the survival of local residents. With the passage of time, the old inhabitants were either killed in ensuing fights or were subjugated. Individuals that still remained were perhaps assimilated with the newcomers through marriage leaving no trace of how they came to an end.

There are no pieces of evidence to support the view that they too, like Cro-Magnons, were assimilated with the newcomers. However, as it happens in war, the women of the losing side are subjected to rape by force or with consent. The disappearance of Neanderthals from the area may have happened because of their loss in wars or through the marriage of newcomers with local women. The local inhabitants were gradually eliminated over a span of 13,000 years or around 27 thousand years ago.

The union, in some cases, may have resulted in genetic mixing between the two giving rise to a new breed of early modern humans in the area. The proof of mixing genes of red colour of hairs found in some modern people with long noses originating from Europe, is supposed to have come from Neanderthals.

The pattern described above has been repeated everywhere around the world where new modern humans went. The old habitants were either killed in war or assimilated through marriage or slavery. The mixing of genes owing to sexual interaction between the early humans and new modern humans had started the process of creation of new races and tribes of varying colours and languages (30: 22).

A part of the lot migrating from Arabia on reaching Fertile Crescent diverted towards east and occupied by 40 Tya (thousand years ago) southern parts of Central Asia. This region became a common point from where many branches spread over time to South-East, East and North Asia.

From that common point some members dared to move into cooler regions northward and around 35 Tya spread over the northeastern parts of Kazakhstan.

Another part continued its journey south-eastward and by 40 thousand years ago reached Myanmar. Some of them continued to move and by 10 thousand years ago had spread over the Indonesian Islands of Java and New Guinea.

A small group broke away from Myanmar 35 Tya and occupied northern parts of Laos and Vietnam.

Some residents of Laos and Vietnam began to disperse northward and by 10 Tya had occupied lands in northeast China.

Streams of early Modern humans migrating towards east and south Asia encountered *Homo erectus* living as far as China and Java. The fossil remains indicate their presence in those areas until 27 thousand years ago. That was perhaps the time when they were overpowered by modern humans and their existence came to an end in these lands.

When Ice Age began to lose its grip, one group left Kazakhstan and moved across the northern mainland of Asia crossing Bering Strait by 20 thousand years ago.

A group turned southward from this common point in Central Asia and moving across Khyber Pass it occupied lands in central parts of India 30 thousand years ago.

Another small group from the original mainstream migrating from Arabia, before reaching Fertile Crescent moved around 25 thousand years ago along the coastal plains of Persian Gulf and the Indian Ocean through Iran and Pakistan and arrived in West India.

Dispersal in Europe

On reaching Central Asia, a breakaway group of migrants from Arabia, instead of continuing their journey onwards to East Asia, turned to the west towards Europe. The group going westward crossed central Europe and settled on lands in Portugal 30 thousand years ago.

While passing through central parts of Europe a small group bifurcated from the mainstream near about Belorussia 10 thousand years ago and turning backwards in southeast direction reached Aral Sea and settled there.

Some other early migrants from Arabia turning westward settled around 45-40 Tya in the Levantine region of Fertile Crescent. A long-settled stay helped the population to increase their number until it crossed the limit of carrying capacity of the area. It led to further migration towards suitable places in southern parts of Europe.

The radiation of groups took place in three streams over different time periods. The first group moved nearly 35 Tya and occupied fertile plains to the north of Italy, the second went by 20 Tya to Greece and Albania, while the third settled 10 thousand years ago in Italy.

By the time cold conditions had waned in Europe, some groups ventured westward from Kazakhstan and moving across northern Europe occupied areas in Finland by 10 Tya.

Dispersal in Africa

Beside dispersal of Perfect Humans through easier routes towards Asia and Europe, there was another route, though troublesome, available to migrating population out of Arabia via Red Sea. The Red Sea, with the exception of a few pools of water, was almost dry at the height of Great Ice Age around 50 to 45 thousand years ago. It facilitated some groups of the descendants of Adam to move out of Arabia by crossing Red

Sea and entering into Africa. This was the same route adopted by their ancestors while coming into Arabia from Africa some 60 to 50 thousand years back.

The first group after crossing the Red Sea moved up to Sudan in East Africa where it settled in Sudan some 35 thousand years ago.

Later on, a few continued their journey northward from Sudan and by 20 Tya spread over the Levantine region.

Some others turned to the southwest and crossed through central Africa reaching western parts of Africa by 20 thousand years ago.

A group of people living in Levant migrated towards North Africa and moved westward along the coast of the Mediterranean Sea spread over Algeria some 10 thousand years ago.

A study published in *Science* 2879 on 8 October 2015 confirms that there was a substantial movement of people from West Eurasia back to Africa. The migrating farmers passing across Europe had spread, possibly about 3500 years ago, over not only northern parts of Africa, but had also reached deeper into the continent.

Dispersal in the New World

It has been suggested that the New World was first inhabited by archaic Homo sapiens like Homo erectus and Neanderthals. These were robust and powerful persons and a group of them had dared to come to America from Siberia by crossing over an ice bridge existing between the two continents during the great Ice Age.

At the end of the glacial period when climatic condition began to lose its grip, a group of Homo sapiens-sapiens or Modern, anatomically perfect humans moving across the northern mainland of Asia between 20-17 Tya, crossed Beringia land bridge and entered America.

Modern Humans. are believed to have moved from Alaska southwards along the Pacific coast and passing through Canada and Central America entered South America and by 14,500 years ago reached Chile and by 10,000 years ago had spread over Paraguay.

On their long journey through the two continents they have left the proof of their presence in the form of fossils/ archaeological records scattered at various places.

Briefly, it may be concluded that the family of perfect humans which started from a single pair of Adam and Eve had spread in 30 to 40 thousand years all over the world. It was the fulfilment of God's words in Quran that He Had turned all the men into nations and tribes (49: 13).

PART-IV

Death and Hereafter

(Stage VII)

CHAPTER X

Universal Death

Death is defined as cessation or termination of life. It may be caused by unnatural events like accidents and predation by other living beings, or it may be natural after going through the process of aging. In nature every living organism born alive begins to grow, increases its size, and then its structure and functions begin to deteriorate leading to old age and ultimately stoppage of every bioactivity in death. The process of growth and aging is therefore an integral part of life terminating in due course into a natural death.

Life in fact is the functioning of all vital organs of the body like respiratory system, circulatory system, nervous system, digestive system, etc. When vital organs stop work, life comes to an end or the body becomes dead. Technically or clinically however the body is said to be dead when brain function finally stops. Death is an irreversible process as no dead can come back to life.

Belief in death differs greatly from religion to religion. Some believe that humans never die, as there is an afterlife for them and they go to heaven or hell. Others believe in reincarnation that the body continues to born again and again in the same form or in some other form like that of animals or birds. Nearly all the religions believe in afterlife, heaven and hell, or soul. Among innumerable beliefs about death, major religions like Christianity, Islam, Hinduism and Buddhism are based on firm ground about their belief in death.

In Christianity, death is more often connected with the life and resurrection of Jesus. Ecclesiastes 9:5 tells that all the living ones know that they shall die and that they will remain in deep sleep without any thoughts, emotions or consciousness. Psalms 146:3-4 say that man dies when his breath goes and he returns to earth that very day his thoughts perish.

Bible explains that there is a time to die when a person goes into a period of dormancy until the second coming of Jesus when his fate would be decided. The salvation from sin is possible only through Jesus Christ who may also bestow the gift of eternal life.

Buddhism regards death, rebirth, and reincarnation as the main purpose of religious life. It is to escape from the cycle of death and rebirth and achieve Nirvana or free oneself from worldly desires.

According to Hinduism, when an individual dies his soul is reborn into a new body. The new body may be that of an insect, animal or a man depending on the good or bad work performed during his lifetime. The cycle of death and rebirth ends only when a soul realizes its true godliness. Such resurrection may take many lifetimes to achieve the goal.

Islam does not consider death an important event as it is just the preparation for movement from present temporary or short life to a permanent life in the Hereafter. This fact is repeatedly and forcefully mentioned at many places in Qur'an. After death, each person will be raised to his or her permanent residence in paradise or hell, depending on his or her actions in this life.For them death is merely a means of transfer of life of performance in existing world to life of judgement and reward in another world.

Death of a man in Quran is considered in two planes. One relates to the death of an individual who is developed in the womb and is born in ontology. The other death considers the formation of man from clay and passing through various stages it enters into its sixth stage called by God 'la Maiyatoon' when the life story of man in this world comes to an end.

Death of Individual

Death is a universal phenomenon and every living organism created by God will have to taste death (3: 185; 29: 57). Humans too have been experiencing death from the day they were created. Death may come to an individual independently or may catch up with a group of individuals, small or large, at the same point of time. The life and death are the sole prerogative of God. These are the only aspects of life where men feel helpless as they have been given no choice to decide about it: (23: 15; 45:26).

If a person takes his own or other person's life he should not think that he is the master of death. He only is the medium for executing the Will of God. The life span of every person has been recorded in His 'Book of Life' and the death will happen on fixed time (3:145). When the predetermined hour comes, death overtakes one breaking every type of barrier erected or medicine provided to protect and save (62: 8; 4: 78).

The time of death as pointed out earlier has already been fixed by God. It may occur according to His plan from the date of birth to a ripe old age (22: 5). Like time, the place where one will die is known only to God: "...nor does anyone know in what land he is to die..." (31: 34).

Man was created by God as a Superior Being and was endowed with faculties of observation, thought, analysis and judgment. He was provided the power of judgment and choice of independent actions to lead a just and good life agreeable to God. The main purpose of God in creating man therefore was to test how far he was successful in carrying out the responsibility and find who is the best in performance (67: 2).

Death of All Life Forms

This is the Sixth Stage in the Journey of Life when in future everyone alive will die or according to Qur'an it will be '*la Maiyetoon*'. At the appointed moment not one, two or a group of individuals, but total race of mankind and every other living organism will be put to death. There will be complete cessation of life (2: 28; 15: 23; 22: 66; 56: 60).

Since the dawn of time, philosophers and thinkers of all ages have attempted to explain the meaning of existence and its ultimate fate. Aztecs of southern Mexico believed that life and even earth is not permanent but has five periods of existence; each represented by a Sun. They thought that the fifth one or the last Sun was destined to end in 52 years by a large earthquake that would tear the Earth apart.

In Indian mythology, the cycle of birth and death of life on earth is the same as their belief in reincarnation that one returns again and again to a new body until he gets salvation. Likewise, the earth ends in cycles through fire and flood, but is reborn.

Ancient Egyptians believed that life is permanent and will never come to an end. God Ra ordered Goddess Hathor to watch evil doers so that there is no chaos and people live peacefully.

Modern day Physicists, Mathematicians, and Cosmologists working on the functioning of the universe have always wondered whether the universe is eternal or not. Or like other matter will come to an end. It has generated fascinating theories in this regard.

Scientists, now, have confirmed that the universe was formed when an infinitely dense point in the abyss exploded with a Big Bang and began to expand outward with accelerating speed. It is now proposed that the universe's gravity will slow the rate of expansion and ultimately stop it. A reverse action known as the Big Crunch will then start leading to contraction and the universe's matter will regress to an infinitesimal point. This theory may be interesting but is unlikely to happen as recent studies reveal that the universe is still expanding with an accelerating speed.

The theory of 'Heat Death' postulates that the gravity is not strong as compared to the rate of expansion with the result the galaxies will drift away from each other leaving all the matter evenly distributed in the Universe. The matter will be there but it will be in the form of a particle.

Aside from substance in the form of stars, planets, meteors, and other objects, all galaxies have supermassive black holes at their centres. These large black holes have extreme gravity and begin to devour adjacent, smaller black holes. This continuous process converts the massive black holes into more massive black holes. However, after sometime the black holes begin to lose their mass because they emit what is called "Hawking radiation ". After the last black hole dies only evenly distributed subatomic Hawking radiation particles are left.

Whatever thoughts of modern scientists and Philosopher may be about the doomsday, it is a matter relating to future that nobody can tell except the Creator Himself. The Divine Books will be of help in understanding the events that are yet to happen.

Christianity and Islam, based on a faith of guidance through Divine books, are clear in the opinion that everything living or non-living will ultimately come to an end. Bible mentions that heavens will disappear with a roar; the elements will be destroyed by fire, and the earth and everything in it will be destroyed (2 Peter 3: 10).

Qur'an, the Divine Book of Islam, mentions the end of the world by turning it into a barren ground (18: 8). Life is present not only on Earth, but God, the Supreme Creator, has already scattered living creatures in many places in the universe (42: 29). If living beings are to be put to death for testing their deeds, extra-terrestrial life scattered in the universe will also face the same fate.

It is not only the earth that will be destroyed, but sun and moon too will be affected. "He has ordained the sun and the moon, each one runs (it's course) for an appointed time." (13: 2). It will be total destruction of sun, moon and earth together with everything therein including man and other life forms. In fact, the entire universe along with extra-terrestrial life will come to an end. It will be a universal phenomenon of total destruction of all the heavenly bodies including billions of galaxies, their stars, planets, and meteors.

The Doomsday

The Doomsday will appear all of a sudden with a catastrophic Big Bang. It will be a tremendous bang which will spread a wave of terror throughout the universe. The earth will be converted into an uninterrupted level stretch with no mountains (27: 87). Whole earth at that time will be in the grip of violent quakes that will shake the entire earth. These powerful earthquakes beginning deep in the earth will throw out all the material present inside it and becomes empty. (99: 1-2; 84: 4).

All stars of the universe will undergo a chain reaction just after the loud bang and will be lost. The whole universe will be affected by such reactions when the stars become dim and collapse, sun and moon are joined, and the sky is stripped away (77: 8; 75: 9; 81: 2; 81: 11).

The end of time ordained for sun is described in Verse 81: 1 as: "When the sun is rounded (swelling like a ball)". The swelling of sun is possibly the result of atomic reactions taking place inside the star. The atomic reactions continuing for a long time predetermined by God will consume all the resources and the sun will ultimately run out of its energy. Faced with this situation the sun will swell to become a rounded red giant. The intense heat given off by this giant red star will cause the oceans or seas on earth to boil as aptly described in Qur'an: "When the oceans boil in a swell." (81: 6). The ocean water on boiling will evaporate leading to extinction of all life on Earth.

Destruction will spread over the whole universe. It will be caused by earthquakes of extremely high intensity, massive volcanic eruptions, and thermonuclear explosions in the solar system and in other heavenly bodies. This will be the period or the Day when the heaven will fold like the folding of a sheet for the records (21: 104). All matter in the universe will be compressed into an infinitesimal point as the heavens roll back or fold. Every living being, wherever in the universe, will perish.

Time of Death to All

There are some indications about the timing of Doomsday in the Divine Books of both Christians and Muslims. Matthews 24: 5-8 mentions Christ as saying: "Many will come in my name, claiming, 'I am the Christ,' and will deceive many." These are only the beginning of birth pains; the actual end is still to come. Apostle Paul discussing last days of the world says that the Spirit has advised people to not abandon their faith and to not follow deceiving spirits and demons in later times (1Timothy 4: 1). However, in 2 Thessalonians 2: 3 it is said that the Day will come only after there is a great rebellion against God.

There is no information in Quran about the time of death to all. It is kept a closely guarded secret and no one knows it except God (20: 15; 7: 187; 79: 44). The only hint given in Quran 1400 years ago is that the end may be very close (33: 63; 21: 1; 40: 59). What to say of arrival, there are some misguided people who are adamant and even reject the possibility of the day of doom.

God has categorically mentioned that the time of appointed End is known only to Him, but it has also been mentioned that He reveals some information to His chosen Messengers (72: 26-27). God did not reveal the time of Doomsday to Prophet Muhammad, but has only informed that it has already started (47: 18).

God has revealed some signs in Qur'an about the end of the world for the guidance of mankind. These are about the events that had occurred before, after or when the Qur'an was being revealed. A most important sign mentioned in Qur'an is the information that Christ will come to Earth before the onset of the Last Day (43: 61). Qur'an also states that all the People of the Book will believe in Jesus before he really dies and will be a witness against them on the Day of Resurrection (4: 159).

God has mentioned the splitting of moon as a proof of the beginning of the Last Day (54: 1). It is a saying common in Muslim world that when Prophet Muhammad pointed towards moon, it split into two. The phenomenon however, is against the law of Nature as any astral body breaking into two pieces cannot re-join together. When in reality, it is a mistake of translating the Arabic word '*shaqqa*' as splitting. The word 'shaqqa', beside split, also means 'ploughing' or 'digging' the earth as it is used in this Verse: "...Then split the earth in clefts." (80: 25-26). Perhaps the moon developed a huge, wide and very long chasm that gave the impression to the viewers from the Earth as if the moon has been broken into two. On the Last Day, the Earth likewise will be shaken violently by powerful earthquakes and will split through innumerable cracks and crevasses (56: 4-5).

A curious sign provided by God about the approach of the Last Day is baffling interpreters of Qur'an for long. It concerns the emergence of an animal of the earth that will be a sign for approaching doomsday. The humans of those days in general will be atheist denying the existence of God. They will refute God's revelations and the Day of Judgement when all will be presented before God to face the consequences of their wrong deeds. The 'Animal of the Earth' will remind them to abandon wrong doings and follow right path to Almighty God "(27: 85).

The Arabic word '*Dabba*' means animal or creature, who feels difficulty in walking. It also refers to a body who under the influence of alcohol feels difficulty in movement and gradually but imperceptibly continues to decay. The qualities of such a creature may be summarized as: It is a moving, living thing, is a product of earth, and speaks and propagates a particular message to the people. The controversy arises from the fact that how '*Dabbatul ardh*' or Animal of the Earth can communicate to all mankind. One theory that fits all of the criteria for 'dabba' is that it is a highly advanced humanoid robot. It has the appearance and behaviour of a living object, moves like an intoxicated body, is constructed of earthly materials, and speaks and conveys a specific message. This possibility may easily be ignored because it is not created by God and talks only those words that had been fed into it by humans. Anyway, God knows best about this creature that will come as a sign of approaching end of the Earth and who will call people to believe in God and on the Day of Judgement, and follow right path to become one of those rewarded by God.

Another important sign mentioned in Qur'an relates to smoke in the sky that will engulf all people. It will be like the smoke that had appeared with a Big Bang when the first God particle was exploded leading to the creation of the heaven when it was smoke (41: 11). In the same way when Universe will collapse with a loud bang it will be smoke (44: 10).

Appearance of smoke in the sky does not indicate approaching end of the Earth, but in fact will manifest when the Day has actually started. It will be the time when massive volcanic eruptions, explosions in the solar system and in other heavenly bodies, and bombardment or pounding by meteorites, all together will fill the atmosphere with smoke and dust (56: 4; 69: 14). The suffocating smoke spreading all over the sky will be a painful torment to the people. It will not be limited to the Earth but, by the Will of God, will cover the whole of the collapsing Universe.

The poisonous and suffocating smoke will be one of the causes of death to all living organisms. Everyone upon Earth will perish except the Face of our Lord, full of Majesty, Bounty and Honour (55: 26-27).

CHAPTER XI

All Raised Up

Afterlife, life after death, or life in the Hereafter, is a concept that relates to a future of mankind unknown to anyone. Human mind gifted with the power of critical assessment has always been busy in figuring out what awaits people once they die. . Those who do not believe in the existence of God reject outright the idea of any afterlife. But God conscious people trying to seek purpose of creation of man on earth, commonly believe in the existence of the Hereafter, where he will live forever based on the nature of his actions performed during his life on the earth.

Cosmologists or Biologists do not talk about any life formation after the end of the Universe. There are, however, some who theorize about the possibility of formation of a new Universe. The hypothetical model known as the Big Bounce implies an oscillatory universe wherein the universe starts expanding with a Big Bang followed by a rapid contraction. It is suggested that the force generated by rapid compression during the Big Crunch is enough to start up a new Big Bang and bounce the matter back. Willem de Sitter et al. (1996) endorsed the concept of Big Bounce, but it is still up for debate.

Mythology and religious philosophies however promote the idea of life continuing after death. The nature of transfer from earthly life to the Hereafter differs between various beliefs. Some believe that the transfer takes place in a spiritual way while for others it is a physical transfer. For some, individuals are born again and again on this earth until the acts are purified and one enters into a spiritual plane. There are also some who think that the individual after death is lost in oblivion and does not remember anything about his earlier life. In Abrahamic tradition the dead are placed into specific places after proper judgement of their lifetime activities.

Ancient Egyptians believed that, death was a temporary interruption of life and the body goes to the Kingdom of Dead. From there those who were pious to the gods and their bodies were properly mummified, will pass on to eternal life, while those who were heavy with evil to gods will be devoured.

The afterlife is also prominent in ancient Greek religion. The souls of the dead remain alive and are placed in the kingdom of the underworld by Greek god Hades. From there they are sent to different places according to their performance in life. For example, those who lived a life of piety are sent to sunny, peaceful and green Elysian Field; the people who remained evil and rebelled against the gods were placed in the Tartarus.

The soul, according to Hinduism, is indestructible and therefore never dies. It leaves the dead body and enters into a new one in a cycle of birth and death. The soul enters into a new body according to the deeds performed by the person in the last birth. If the person had performed bad deeds the rebirth would be in the form of lower animals, while in case the performance was good, he is reborn in a human form in a good family. Persons who had accumulated a set number of good deeds acquire salvation and stay finally with supreme God for ever.

Death, in Buddhism, is simply a break from this materialistic world. The soul of a dying person achieves awakening and a peaceful state of mind. Like Hinduism, Buddhists too believe in reincarnation wherein the dead person is reborn to a new life on the earth. The state of new birth into lower or upper level depends on the morals of the dying person. The death and reincarnation continue until the deceased finds the bright light of wisdom leading to topmost level, becomes Buddha, leaves the cycle of reincarnation, and settles in the Pure Land.

Almost all Abrahamic religions fundamentally believe in life after death and the final judgement. The nature of resurrection however interpreted by adherents has introduced variations according to differences in their beliefs. Orthodox Jews believe in the raising up of physical body but modern Jews reject the possibility of literal survival after death. Christians believe that a spiritual body different from our current body is resurrected after death. Many more however continue to believe in the basic tenets that earthly bodies are raised up after death. Adherents of Islam generally agree that physical body will be resurrected on the Day of Judgement. Still a small minority interprets Qur'anic message differently and holds a dualistic view of soul or body resurrection.

Hereafter

The subject of life after death relates to a far off future that can never be seen, witnessed, or be proved scientifically. One has to turn towards the Creator to know what will happen to His creations after the existing life comes to an end. All the Divine Revelations like Torah, the Psalms, the Gospel and the Qur'an speak about resurrection of life after it ends in this Universe.

Qur'an is the final Revelation and this Book of Truth presents detailed information about what lies beyond death and what will be the future of mankind. Everything revealed in Qur'an is available word by word in its original form and none can change it (18: 27). Qur'an is the Book that carries true words of God. Any information about life in the Hereafter can be obtained without any doubt from this Book. This information is from God Who knows the future and the dead will acknowledge it on the day of Resurrection (27: 65).

Qur'an forcefully informs about death not being the end of life formations, but that it is He Who has given life, will resurrect them too in the Hereafter (10: 56; 6: 38; 19: 40; 36: 79).

It is also pointed out that majority of humankind who do not believe in God and His Messages deny the possibility of dead rising again (11: 7). Some others will say that when they have turned into dust and stones after death how can they and their forefathers be raised (37: 16-17; 27: 67).

There are verses in Qur'an that try to explain by way of similitude between rain and earth as to how God brings the dead matter to life. People are asked to see when rain is poured on their dead barren land, it is stirred to life after its lifelessness and produces many types of plants (22: 5-6; 30: 24; 43:11). It is He Who sends rain-laden clouds to a dead land making it raise every kind of produce. Likewise, He will raise up the dead (7: 57; 35: 9; 50: 11). Indeed He gives life to dead earth and thus shall raise you alive (30: 19; 36: 12).

Everyone will be raised up to face the consequences of their performance. Even those who had strongly sworn in in the name of God, an oath binding upon Him, that they will not be resurrected, will nevertheless be raised up by God to show the truth of the revelation about the Hereafter (64: 7; 23: 16).

Day of Resurrection

A new Universe will have to be created that could accommodate a multitude of people rising from the dead state. A trumpet call with a mighty blast will start its creation. It will be a 'Big Bounce', the name given by scientists for re-creation of Universe from the singular point to which the original universe would have reverted. The re-creation of Universe had already been told by God 1500 years ago when it was mentioned that indeed it will be repeated (Big Bounce) like the first reaction (21: 104).

The old earth and the universe will be substituted by a totally new formation. The earth will be flattened free of any obstruction blocking the view and appear like a prominent plain (14: 48; 18: 47).

When the new earth is ready to receive its inhabitants there will be another call with a loud shout heralding the onset of the Day of Resurrection (50: 41-42). The blast will shake out all from their slumber. All those in graves in heavens and earth will rush out in hordes from all corners through numerous gateways opened in the universe (78: 17-19; 36: 51, 53).

Everyone will be seized with fear except those whom God will exempt. They will ask each other why have they been raised up from a resting place, what is happening, and what is the matter. Then they will realise it is the Day of Resurrection (27: 87; 36: 52; 37: 19-20).

Those who do not believe in resurrection will on rising wrongly think that they had remained therein, (in their earthly existence), for an hour or half a day. But those who had knowledge and were faithful will be told that they had remained within the extent of God's Decree till the Day of Resurrection (79: 46; 30: 55-56).

This Day will be harsh for non-believers as they will be disregarded by God the same way they had denied the Messages. Those who turn away from the Messages shall be raised up blind on the Day of Judgment since they too were blind towards God's messages and disregarded His signs (20: 124, 127; 17: 97).

All of those gathered will be terrified, forlorn and self-concerned. Every relationship will cease to exist and even close ones will not speak to each other. Each person will have enough concerns of his own and will not care a bit for others. This day a man shall flee from his own brother, and from his mother and his father, and from his wife and his children. (23: 101; 80: 34-36).

The horrified multitude assembled on this Day will be looking forward to know their fate. The main purpose of creation was to test how far man remains obedient to his Creator. The objective will not be achieved if the process of creation stops with death, but it is to be concluded after proper assessment of the work done during their lifetime.

Scrolls of Deeds

(Nama-e A'maal)

The judgement of rewards or punishments will purely be based on ones own performance (21: 47). God had already reminded men and jinns that He did send Messengers warning about meeting you on this Day and to avoid idolatry and sin, and believe in life after death but some rejected Faith (6: 130; 78: 40).

Those who do not follow the Command of God or deny His words shall face dire consequences on the Day of Judgment. Persons who considered some pious dead men as partners of God and worshiped them will be asked on the fateful Day to invoke them but will get no reply, and will fall in the fire where from there will be no way of escape (18: 52-53).

The extreme act of rebellion against God is concealment of Revelations in the Divine Book. People committing this act for little profit will be neglected by Him and will receive grievous punishment. There may be small worldly gain for people who hide God's revelations in the Book but indeed in the life after death they will fill their bellies with fire. They will have a painful punishment and God will not speak to them on the Day of Resurrection (2: 174).

All deeds performed by the individual during his lifetime are recorded in a scroll known as *Nama-e A'maal* or Scroll of Deeds. The recording is done by angels who are kind and honourable, know all and everything you have done, and everything small and big is written down (82: 10-12; 54: 52-53).

Every action performed by an individual is recorded without any change (50: 29). Everything uttered or done is recorded In an open Book.

Day of Testimony

On the Day of Testimony all will be arranged in rows for presentation before God. Every soul will be accompanied by a driver angel carrying the record of deeds performed by the soul and with another angel to witness the acts performed. The Throne of God held by eight angels on sides will be placed above the assembled multitude (18: 48; 50: 21, 23; 69: 17). All those raised up, gathered and assembled before the Lord will anxiously be waiting to see the result of their actions performed during their earthly life (42: 29; 45: 26; 32: 11). The record of deeds will be handed over to respective person. He will be asked to read and see how his record is sufficient an account against himself (36: 12; 17: 13-14).

The proceedings on the Day of Testimony will be sober and well organized. No one will dare to speak except those specially allowed by God and those too will only witness the truth (76: 38; 11: 105).

Record of deeds will be opened and everyone will be surprised to see that nothing is hidden in the records. Any word spoken or act performed none whether small or great is missing in the record (18: 49). Every individual will in fact find himself fully exposed before God. Even the secrets they never revealed to anyone are present there (6: 60; 69: 18).

All are presented to God for deciding their fate. He being an embodiment of Judgement and impartiality shall make a just decision about them (6: 62).

The Judgements

At the time of judgement, the Scrolls of Deeds will be laid open for every individual to see what he did and know his fate (81: 10-14; 58: 6).

How the record will be shown is explained by God with an example. It will be like the actions performed during the day are put on record and are known the next day after a gap of night's inactivity. Likewise, the deeds of a person during his lifetime are well recorded and will be shown to him on resurrection after death (6: 60).

It is pointed out that as people swear to Prophet Muhammad about their actions, likewise on seeing their record of deeds on the Day of Judgement some will deny before God their actions performed during their lifetime. Facing this doggedness God will seal their tongues. Their mouths and hands will start speaking, and their feet will witness of what they had done (36: 65; 24: 24). When the enemies of God are gathered on the Day their hearing, their eyes, and their skins will testify against them of what they used to do. The skins will say that God has made everything to speak (41: 19-22).

Both good and bad deeds will be put in a balance to show the individual his fate. If the good deeds are heavy in scales and greater than bad deeds, the person will be handed over the judgment in the right hand and he will turn to his people, rejoicing. On the contrary, the recipient whose records are bad and light and receives it on the left hand will be dejected and thrown into Hell (84: 7-12; 23: 102-103). Faces of believers will be shining and beautiful while those of non-believers will be sad and dejected (75: 22-24; 80: 38-42).

The extreme act of rebellion against God is concealment of Revelations in the Divine Book. People committing this act for little profit will be neglected by Him and will receive grievous punishment. There may be small worldly gain for people who hide God's revelations in the Book but indeed in the life after death they will fill their bellies with fire. They will have a painful punishment and God will not speak to them on the Day of Resurrection (2: 174).

Those who believe in the Power of God, remain steadfast in their belief, motivate people towards right path, and always seek mercy and peace from Him, will be welcomed by Angels who console them not to fear or grieve but they will be rewarded by the gift of Paradise. In fact these are better placed as they call people to follow God, do righteousness and proclaim to be Muslims (41: 30, 33).

The Hell

Hell was made ready, with its horrible nature of extreme heat and paralyzing cold, to receive each and every transgressor who strongly denied the Revelations of God. The residents of Hell will dwell there for ages undergoing a punishment proportional to their evil deeds as recorded in the Book. There will be no reprieve in the punishment and there will be an ever-increasing torment: (78: 21-26, 30).

The Hell will be filled with a large number of people derived from every walk of life and from every period of time. They will be put to constant torture of every imaginable nature because of their obstinacy in denying the Commands of God and the establishment of the Day of Judgement (25: 11; 36: 64).

Persons who arrogantly deny the Power and Superiority of God are the worst victims. Although they acquiesced to the decision patiently, they will not be granted any favour for amelioration of the punishment and fire will be their home. Moreover, they will be accompanied by those who made sin alluring to them and by the previous generations of jinns and men who were destroyed long ago for their wrongs.

Punishment will be severe also for those who arrogantly disbelieve in the Revelations of God. The unbelievers used to indulge in vain talks and other irrelevant activities to prevent and disturb Quran listening during the discourses by the Messenger of God.

They will be rushed to Hell with their eyes lowered in dejection and with ignominy (70: 42-44). They will be severely punished for rejecting Verses of God. They will have to taste for ever an intense Fire (41: 26-28). They will not be allowed entry into the gates of Paradise. Instead, the rewards for the wrongdoers will be bed of fire in Hell (7: 40-41).

People filled with pride who scorned the call of Messengers to worship no god but God went one step further in hate to call the Prophet a mad poet. For this insolence they shall indeed taste the painful doom. It will be a retribution for their rude and disrespectful behaviour (37: 35-39).

The disbelievers of Verses will be driven into Fire.Their faces will be covered by fire and their garments too in the Hell will be made of fire. In intense heat their skin will be roasted but the same will be replaced as quickly as it is burnt. That is the way of God to let them taste the punishment (14: 50; 22: 19; 4: 56).

Non-believing couples while being led towards Hell will be quarrelling and blaming each other for misleading from the right path. But finally they acquiesced to their fate and admit that they themselves were at fault. That Day they will share the punishment and will be seized, dragged and thrown into an eternally fuelled Fire. The torment of boiling water will be poured upon their heads (37: 22-23; 44: 47-48). Every time they wish to get out from such anguish, they will be returned to taste the punishment of the burning Fire (22: 21-22).

The tree of Zaqqoom has its roots in the bottom of Hell and its fruit-stalks resemble heads of the devil. The inhabitants of Hell will eat its fruit and quench their thirst by drinking a mixture of fruit and boiling water (37: 62-68).

Finding the torment unbearable, the denizens of Hell will approach its Keepers and the Chief Keeper to lighten their punishment. On getting a stern denial, they will dare to seek mercy from God to lighten the torment for a day. The request would be rejected and they will be told to remain there and never ask for any reprieve. (40: 49; 43: 77; 23: 107-108). The Hell-dwellers will be curtained off and will never be allowed to view Paradise (83: 15).

The Paradise

Entry into Paradise is not any special favour but is a reward of God for person's own performance with respect to his faith and belief. After final judgement all those successful will be handed over their records on the right hand. Getting the record on the right hand is an indication of their acceptance by God.

They will be overjoyed with the thought that the life thereafter will be pleasant. Every soul who is made to enter Paradise will feel proud for the grand achievement (69: 19-21; 3: 185). The righteous and a reassured soul will be asked to return to his Lord well contented and well-pleasing unto Him. It will be a fulfillment of their heart's desires (89: 27-28; 78: 31).

The believing men and believing women will proceed on a brightened path towards the Paradise. On the way they will hear good tidings that they are going to a high garden full of fruits and beneath which rivers flow. They will eat, drink and enjoy their everlasting stay with full satisfaction (57: 12; 3: 15; 69: 24).

Those authorised to dwell in Paradise will proceed towards it rejoicing and singing the Glory and Praise to God. On arriving there the gate-keepers will open the gates addressing them by welcome words of Peace be upon you, and ask them to enter the Garden and join others in gaiety. They will Praise God for fulfilling His Promise to them and permitting them to dwell in the Garden. It was certainly an excellent reward to righteous workers. The Decision between them at the Day of Judgment was perfect justice, and the cry of Praise on all sides will be raised to God, the Lord of the Worlds (39: 73-75).

Paradise will be occupied by a greater number of people coming from early days of human history and a few of those coming at later times (56: 13-14). They will be placed in different ranks as assessed according to the degree of piety of their deeds. Whoever is a believer and has done righteous and good deeds will get high ranks. Those with unflinching faith in the Oneness of God will occupy foremost place near God in the Gardens of Delight (20: 75; 56: 10-12).

Sincere and devoted believers will be honoured in the Gardens of delight. They will be relaxing on couches facing each other and enjoying bounties available there. They will be among thorn less lote-trees and clustered plantains, Fruits of every kind will always be present near at hand. A cup of crystal-white drink will be brought to them from a gushing spring. It will taste delicious, will be free from bad effects, and none will be intoxicated (56: 28-32; 37: 40-47).

The occupants of Paradise, beside other enjoyments, will be bestowed with the company of special creations of God that will be of modest gaze with well-protected lovely eyes, virgin, and of their own age (37: 48-49; 56: 35-37).

The greatest delight for the residents of Paradise will however be the fact that they will be blessed with the everlasting view of their beloved God (75: 22-23).

Epilogue

A highly condensed, compact and miniscule mass of gases (heavens) and matter (earths) was exploded with great creative power of Almighty God. The universe thus created was also made to expand, a process that is still going on. These qualities of the Big Bang are revealed by God Himself: "We have built the heaven with might, and indeed, We are (its) expanders." (51: 47).

The formation of Universe was the first step towards providing a playground for the living organisms. The Big Bang occurred 14 billion years ago and the space expanded to its present size. It includes in its fold gases, innumerable galaxies, galaxy clusters, stars, planets, and a variety of matters of different size.

The next step of God in the general plan of creation was to inject life into the lifeless matter of the Universe. In the vastness of the Universe with billions of stars and planets, God mentions that planet Earth was the first to be prepared for habitation by living organisms.

The heaven nearest to Earth with its sun, earth and moon is an independent, closely linked functional unit. The bond is maintained by the rising and setting activity of the sun. When sun rises over the earth it shines bright and on setting becomes dark or invisible. This unit or system is called in scientific parlance a solar system that, besides a sun, an earth and a moon, also includes other planets and moons revolving around the sun. This close-knit solar system, a small part of the galaxy called Milky Way, is referred to by God as the 'lowest heaven'.

God mentions setting up six more heavens above the first heaven (2: 29). The description is never complete without mentioning the presence in each heaven and earth that is similar to planet Earth of the lowest heaven: "Allah is He Who has created seven heavens, and of the earth..." (65: 12). The preceding Verse indicates that like the seven heavens God has also created earths in similar numbers, i.e. seven. These seven heavens with seven earths from the lowest on one end to the highest at the other make a full-fledged Universe.

All the seven heavens in the Universe where their sun's rising every day controls life are in themselves whole worlds or '*Al-aalameen*'. He is God, Lord of the seven worlds or seven heavens and the earth and all that is between them. and is Lord of the sun's risings (37: 5). The word 'risings' indicates the presence of suns, one each in the seven solar systems in the seven heavens. It is clarified in a verse mentioning that He is Lord of the two (plural) Easts and Lord of the two (plural) Wests (55: 17). The use of two Easts and two Wests stands for their plural status.

Life in the Universe

Since water and clay in the words of God as mentioned earlier are essential for creation of life, the availability of the two in the first instance was ensured for appearance of life. When water and clay were made available abundantly in planet Earth, it was the place where life dependent on water and clay first appeared.

Existence of extra-terrestrial life is still a matter of discussion in scientific circles, but God specifically mentions that He has created a variety of living creatures (*dabbatin*) and dispersed them throughout the heavens (seven) and (in each) Earth (42: 29).

The word "dabbatin" here refers to walking creatures like us men that are present on planet Earth. At another place in Qur'an the same thing is confirmed for all the heavens that whatever creatures (dabbatin) are present on the earth, and the angels as well, all prostrate to God. (16: 49).

Humans among all God's creations are the only living beings bestowed with knowledge and a free will to act as they like. God's creatures are present in every heaven or all over the worlds (Alaalameen) and can understand and follow the instructions sent to them by Him. The commandments come down slowly whenever required. Everyone knows that God is in knowledge of and is able to do all the things (65: 12).

We do not know the anatomy and morphology of manlike or other beings that have been created successively in other six earths of the Universe. Do they look like us or have different shapes and how do they function in their daily life? God informs that man has been created weaker in strength as compared to others who are stronger. His weakness is because of his creation from sticky clay or wet earth (37: 11; 23: 12).

Creation of man from sticky clay indicates that formations in the other earths being stronger in nature may have been created from dry earth or from some other matter. Creation from dry matter shows paucity of water that may have created problems in the formation of life. Under such conditions, according to God, manlike creations in other places are stronger than those weaker ones formed from wet earth.

Planet Earth lying in the lowest Heaven amongst the seven created by God has the privilege of being the first place where Man, though weak in strength, was the first to be created in the whole of the Universe. Man was not only the first amongst all creations, he was formed in a better mould (95: 4). In recognition of his better form and superiority over other creatures, God made him His Vicegerent (2: 30).

Journey of Life on Earth

The story of man starts from the day planet Earth, where the drama of life was to be played, was fully prepared to get its main actors. The preparation refers to make available all those items that are essential for the constitution of a living body. The earth therefore got ready with ample supply of water and clay that, as God points out, are basic requirements for the constitution and survival of living organisms. Earth got ready with water (oxygen, carbon, hydrogen, nitrogen) and clay with (calcium, phosphorus, potassium, sulfur, sodium, chlorine, and magnesium).

The whole story of mankind after its first step to the last status can be divided into three distinct stages:

1. Day I: Life on Earth
2. Death of Life or Night of Calm
3. Day II: Resurrection (Raising the Dead) or Revival of Life

The three different stages are distinctly explained in various places in Qur'an. The first and the third are referred to as Day and the second in between is called Night. The first one relates to the creation of man and how he fared in his activities until the death of all life on the earth. Next two stages are in the domain of God Who informs in Qur'an that after death everyone will remain in tranquillity for a certain period in a state known as '*barzakh*' (curtain or partition) after the Day of Life on the earth until all will be resurrected on the Day of Judgment.

Day I: Life on Earth

The Journey of Life started 2100 million years ago and after passing through many modifications in anatomy and morphology until 50,000 years ago when perfect man was created. The ideal man, complete in all respects including his own decision-making qualities, was thus created in 2099.5 million years.

The lifespan of humankind on this earth is not known. The Creator in His revelations has informed that the whole Universe with all the living creatures will one day be destroyed. People did not believe in the revelation about the Doomsday but God emphatically said that the Hour will surely come and there is no doubt about it (40: 59).

No one knows when the calamity will take place. The matter is kept secret by God. It has however been mentioned in Qur'an that the Hour is approaching quickly and will appear suddenly (33: 63; 21: 1; 26: 202).

What to say of near-future-approach of the Doomsday, God has also mentioned that it has already started as indicated by certain signs mentioned by Him in Qur'an. The clear signs as described earlier relate to the revelation of Qur'an in the Blessed Night, the splitting of the moon, arrival of Jesus Christ, appearance of a beast of the Earth, a cloud of suffocating smoke that will envelope entire humanity, and some horrible earthquakes (54: 1-2; 43: 61; 27: 82; 44: 10; 99: 1-3).

Death of Life or Night of Calm

The two periods of life and activities on the earth and that in the Hereafter are regarded by God as two days and the intervening period is the night of Death. As the Night of Dead and the Hereafter are in the hands of God no one can tell when and where the two will take place. God has repeatedly informed us that the timings of the Last Day when everyone will die and everything will be destroyed are a secret with Him and no one can foretell it.

An indication of the time of the onset of Night of Calm is however mentioned by God to be when it is said that Quran was revealed in the Blessed Night (44: 3). Further, the occasion of splitting of moon is given as an example of the arrival of the Final Hour in the near future (54: 1). The splitting of the moon is an indication that Last Hour Quakes heralding the onset of Doomsday have already started and their impact was seen on the moon.

Splitting was a physical reality as it was witnessed by locals and by people living at far off places. Possibly the impact of earthquake created a long chasm in the moon that was wide enough to be seen from the earth as if moon has been broken into two parts. Prophet Muhammad has informed that three more earthquakes of great intensity will take place before the final 'Earthquake of the Hour'. Of these, one will occur in due course of time in the East, the second in the West, and the third in Arabia.

With the revelation of first five Verses of Qur'an on this Night, Prophethood was sanctioned upon Muhammad, the Messenger of God. It means that the Day of Doom had already started during the lifetime of Prophet Muhammad. In the beginning first five verses were revealed during this Night and the rest of Qur'an continued to be revealed as and when required over a period of next twenty three years. These two acts of God, i.e. grant of Prophethood and revelation of Qur'an are a proof of importance and of a high position held by this night known as Laylatul-Qadr or Night of Power, Night of Destiny, or Night of Decree.

The Night of Decree had started during the lifetime of Prophet Muhammad but even after 1500 years the actual Night is not yet visible. May be in the system of God's Day and Night, we are passing through a period that in Earth's system of day and night is known as twilight when the night overpowers day and afterwards real darkness of the night follows.

As far as what is going to happen after the Night of Dead actually begins is given in Chapter/Sura 84, Al-Inshiqaq (The Splitting Asunder). It describes 'stage by stage' the events that would take place after 'the sky is ruptured' and 'the earth is leveled out'. God takes oath of The Night and its twilight glow in view of the importance of what will happen during that period (84: 15-17).

It is well-known that in the animal world, according to Recapitulation Theory or the evolutionary history or phylogeny, the race of the animal is repeated in its ontogeny or in its own formation from fertilization to actual birth of the baby. Likewise, the phylogeny or evolutionary history of human journey that would take place after death is briefly described as it's ontogeny during Shabe Qadr in Surah 97, Al-Qadr: "We revealed it (the Quran) on the Night of Power (Destiny)." (97: 1). Verse 4 of another Chapter/Surah 44, Al-Dukhan (The Smoke) also describes that every event or every precise matter will be sorted out on this Night.

Day II: Resurrection and New Life

God had made a promise and that too binding upon Himself that after death He will resurrect everyone to life (21: 104). The resurrection is not difficult for Him as He is most powerful and can gather all of those He had scattered in the heavens and the earth (42: 29).

The Day of Resurrection and Judgment is a reality but no one except God knows it's appointed time and only He Himself can reveal when it will take place. Quran however enlightens that the 'Night of Power' will come to an end with the break of dawn heralding the arrival of a 'New Day' (97: 5).

The Day will start with a Mighty Blast when earth will rapidly expel alive its dead that will assemble expectantly to know the cause of their gathering. Everyone will be assembled on a new earth in a new Universe (14: 48).

The Day of Resurrection is also the Day of Recompense as everyone will be judged according to his performance. The gathering will anxiously be waiting to know their fate. Those with good deeds will proceed to Paradise as against others who will be directed towards Hell. Since it is the Last Day, there will be no more days in future and humankind will live eternally in Paradise or Hell.

God has put His stamp of perfection and finality on the story of life by calling the Last Day as the Day of Eternity. There are many Verses in Qur'an that mention the fact that members of Paradise or Hell will reside there forever and there will be no way of coming out from their abode. Those who are truthful and faithful in their belief in God will remain eternally in Paradise and will never be asked to leave it. Those who disbelieve in the Oneness of God, disobey His Commands and do not follow His Messengers will be pushed into Hell to suffer its torments forever.

The story of life revolves through the formation and expansion of the Universe until its collapse. It will be followed by, as mentioned by God, a long period of calm. Thereafter a new Universe will come into being, life will be resurrected, and will continue therein till eternity.

Acknowledgements

I had never thought even in my dreams that writing an acknowledgment will be a time consuming and tough job. A flood of those who helped me during preparation of this book checked my decision as to who should be left out. But I had to do it albeit with a guilty heart in view of the constraints in this regard.

The first preference should be given to my family members, friends and colleagues who developed my thought process and always extended a helping hand whenever required. My eldest brother, Mohyiddin Siddiqi, was the one who sparked my imagination to ponder over creation of life by God in seven stages. He pointed to a key word 'tsumma' (then afterwards) that separates seven stages as revealed in Qur'an. From that day I started thinking about this information and was able to ultimately develop it in the form of a book. I place my other brother Agha Kamal Siddiqi too in this category for his timely guidance with new ideas and critical interpretation of the Verses relating to life formations. I am greatly indebted to both of them for leading me to a faultless conclusion.

The invaluable help rendered to me by my brother Dr. M. Rafiq Siddiqi, nearest to me in the line of seven brothers, can never be forgotten. Being a D. Sc. in Zoology and a doyen of nematode taxonomy, he was recognised as one of the greatest Nematologists of the 20th and early 21st centuries. He excelled in the study of primitive plant parasitic nematodes that were created in the Devonian period or during the 3rd stage in the Journey of Life. He had also translated Qur'an from Arabic into English. I am proud to have the privilege of getting the benefit of his knowledge of both Science and Qur'an. In fact he appreciated my work, critically examined the manuscript, brushed up the matter, provided new ideas and readily offered useful suggestions. His daughter, Safia F. Siddiqi, was always helpful in managing my correspondence to her father and removing any error in the manuscript. I am greatly indebted and thankful to both of them.

It pains me to write about Dr. Charles H. Southwick who could not see my book in a published form. He was no less than a brother to me and like my other brothers is no more with us. I pray God to rest the souls of all of them in peace. Southwick, renowned ecologist and primatologist, formerly of Colorado University, joined Aligarh Muslim University as a Fulbright Fellow in late 1950's. With my basic training in Geography, I started collaborating with him in the study of population growth and geographical distribution of primates specifically in and around Aligarh and in general in North India. I am thankful to him for providing me an opportunity to study and understand the behaviour of primates who gave rise to two-legged animals that after refinements and re-shaping developed into perfect humans. I gratefully acknowledge his help and support extended to me at every stage of writing the manuscript.

I feel great pleasure in acknowledging the support of my nephew Dr. Tariq Murtaza Qadri Professor of Physical Health and Education for constantly coaxing me to continue this project and his curiosity to know more often about the progress of it was of great encouragement and a boost to my spirits. His appreciation of what is being done in my view was an incentive for me to accomplish my goal.

At this point I also feel my duty to mention the help extended to me by Mrs. Shabana Tabreiz in providing basic reference material about the origin, anatomy and physiology of various animals. I thankfully acknowledge her valuable support in this regard. Dr. Tabreiz Ahmad Khan, Professor of Botany, deserves my heartfelt gratitude and thanks for extricating me from any dilemma that faced me in the process of writing the Book.

I will feel guilty if I do not express my feelings imprinted deep in my heart about my close and loved ones who always stood with me in spite of their own engagements. My wife, Shakira Farooq remained a mute sympathiser and an embodiment of patience bearing long periods of neglect from me during my involvement in preparing the manuscript. My special thanks are always extended to her. My daughters, Mumtaz Jabeen and Zarreen Farooq, never failed to offer expert comments and fruitful suggestions. Mumtaz Jabeen, Ph.D. In Zoology with specialisation in Genetics, and second author of the Book, was responsible for writing most of the preliminary Chapters and helping in refining others. Zarreen Farooq, a Postgraduate in Biochemistry, readily extended valuable and befitting suggestions. My sons-in-law, Dr. Azmat Ali Khan, Associate Professor of Biotechnology at King Saud University and Mr. M. Kamran Naim Khan, Master in Business Management and CEO in a Company at Jeddah never hesitated in offering critical opinion and invaluable suggestions about my writings. Moreover, they put their computer facilities at my disposal for long hours at the cost of their own time loss. In fact major part of the book was finalised during this period when I visited them in Saudi Arabia. I will keep for ever in my heart the sacred feeling of love I received from all of my loved ones and pray God to help them succeed in every aspect of their life.

Looking towards those of the Siddiqi family spread over far and wide in the world, I found them ever anxious to offer their appreciations and best wishes for successful completion of my work. I am thankful to all of them, for expressing, by way of word or with personal contact, useful suggestions and wish them every success in life.

I wish to extend my grateful thanks to all those known and unknown persons who had directly or indirectly helped me in writing about the Journey of Life. The foremost of all or in fact who should be the first to be acknowledged is Dr. R. Southwood known to me through his remarkable book, The

Story of Life. It was the main source of inspiration and provocation for me to put forth my views in the light of Qur'an. I feel proud to use some of his information especially relating to fossil finds to support my text. Had Dr. Southwood been a believer in the message of Quran, I am sure, he may have presented the facts in a better way than me. I sincerely acknowledge his useful contribution to this field of Biology.

This book may not have seen the light of the day unless unknown helpers have not come forward to support me. I gratefully acknowledge with sincere thanks the cooperation of Editors, Publishers/Printers and all the others involved in the process of printing who helped me in getting the manuscript published.

M. Farooq Siddiqi

Aligarh

List Of Figures

Qur'an Verses Supporting The Text

1: 2

2: 28, 2: 29, 2: 30, 2: 31, 2: 32, 2: 33, 2: 35, 2: 36, 2: 37, 2: 57, 2: 65, 2: 66, 2: 117, 2: 164, 2: 174, 2: 175, 2: 176, 2: 198, 2: 259, 2: 260

3: 45, 3: 46, 3: 47, 3: 54, 3: 59, 3: 77, 3: 78, 3: 96, 3: 145, 3: 156, 3: 185, 3: 186

4: 1, 4: 56, 4: 78, 4: 82, 4: 157, 4: 158, 4: 159, 4: 168, 4: 169

5: 19, 5: 34, 5: 37, 5: 63

6: 1, 6: 2, 6: 38, 6: 59, 6: 60, 6: 115, 6:122, 6:130, 6: 131, 6: 165

7: 11, 7: 19, 7: 20, 7: 22, 7: 26, 7: 29, 7: 36, 7: 40, 7: 41, 7: 50, 7: 54, 7: 57
7: 163 7: 165 7: 166 7: 187

9: 116

10: 31 10: 62 10: 63 10: 64 10: 92
11: 7, 11: 98, 11: 99, 11: 103, 11: 104, 11: 105, 11:106, 11: 107, 11: 108

13: 2, 13: 3, 13: 19, 13: 26, 13: 39, 13: 42

14: 32, 14: 33, 14: 34, 14: 42, 14: 43, 14: 48, 14: 50

15: 9, 15: 19, 15: 20, 15: 21, 15: 23, 15:26, 15: 28, 15: 29, 15: 43, 15: 48, 15: 85, 15: 86

16: 5, 16: 6, 16: 7, 16: 8, 16: 10, 16: 11, 16: 12, 16: 13, 16: 14, 16: 15, 16: 16, 16: 18, 16: 38
16:39, 16: 40, 16: 49, 16: 66, 16: 67, 16: 69

17: 49, 17: 50, 17: 51, 17: 66, 17: 97, 17: 98

18: 7, 18: 8, 18: 17, 18: 27, 18: 37, 18: 47, 18: 48, 18: 49, 18: 52, 18: 53

19: 27, 19: 28, 19: 29, 19: 30, 19: 31, 19: 32, 19: 33, 19: 37, 19: 38, 19: 39, 19: 40, 19: 56 19: 67, 19: 93, 19: 94

20: 53, 20: 75, 20: 104, 20: 118, 20: 119, 20: 120, 20: 121, 20: 123, 20: 124, 20: 125, 20: 126 20:127

21: 1, 21: 3, 21: 30, 21: 31, 21: 34, 21: 35, 21: 47, 21: 104

22: 5, 22: 6, 22: 7, 22: 19, 22: 21, 22: 22, 22:29, 22: 46, 22: 47, 22: 66, 22: 70

23: 12, 23: 13, 23: 14, 23: 15, 23: 16, 23: 17, 23: 78, 23: 101, 23: 102, 23: 103, 23:106, 23: 107, 23: 108, 23: 112, 23: 113, 23: 114

24: 40, 24: 45

25: 11, 25: 12, 25: 54

26: 23, 26: 24, 26: 27, 26: 28, 26: 33
27: 65, 27: 66, 27: 67, 27: 71 27: 72, 27: 82, 27: 83, 27: 84, 27: 85, 27: 87

28: 68, 28: 71, 28: 72

29: 19, 29: 20, 29: 55, 29: 57, 29: 63

30: 2, 30: 3, 30: 4, 30: 15, 30: 19, 30: 20, 30: 22, 30: 24, 30: 48, 30: 50, 30: 55, 30: 56

31: 10, 31: 20, 31: 34

32: 5, 32: 7, 32: 8, 32: 9, 32: 10, 32: 11, 32: 14, 32: 17

33: 63, 33: 64, 33: 65

34: 33

35: 9, 35: 11, 35: 12, 35: 27, 35: 28

36: 12, 36: 33, 36: 34, 36: 36, 36: 40, 36: 51, 36: 52, 36: 62, 36: 63, 36: 64, 36: 65, 36: 71 36:77, 36: 78, 36: 79, 36: 82

37: 4, 37: 5, 37: 6, 37: 11, 37: 16, 37: 17, 37: 18, 37: 19, 37: 20, 37: 21, 37: 22, 37: 23, 37: 24, 37: 25, 37: 26, 37: 27, 37: 28, 37: 29, 37: 30, 37: 31, 37: 32, 37: 33, 37: 34, 37: 35 37: 36 37: 37 37: 38 37: 39 37 40 37: 41 37: 42 37: 43 37: 44 37: 45 37: 46 37: 47 37: 48 37:49, 37: 53, 37: 62, 37: 63, 37: 64, 37: 65, 37: 66, 37: 67, 37: 68

38: 71, 38: 72

39: 6, 39: 66, 39: 67, 39: 68, 39: 69, 39: 70, 39:73, 39: 74, 39: 75

40: 49, 40: 50, 40: 57, 40: 59, 40: 61, 40: 64, 40: 67, 40: 71

41: 9, 41: 10, 41: 11, 41: 12, 41: 19, 41: 20, 41:21, 41: 22, 41: 23, 41: 24, 41: 25, 41: 26, 41: 27, 41: 28, 41: 29, 41: 30, 41: 31, 41: 32, 41: 33

42: 29

43: 10, 43: 11, 43: 12, 43: 13, 43: 61, 43: 71, 43: 77, 43: 161

44: 2, 44: 3, 44: 4, 44: 5, 44: 6, 44: 10, 44: 11, 44: 12, 44: 47, 44: 48

84: 3, 84: 4, 84: 5, 84: 7, 84: 8, 84: 9, 84: 10, 84: 11, 84: 12, 84: 15, 84: 16, 84: 17, 84: 19 84: 20, 84: 22, 84: 23, 84: 24, 84: 25

85: 13, 85: 15, 85:16, 85: 21, 85: 22

86:12, 86: 13

87: 3, 87: 12

89: 21, 89: 27, :89: 28, 89: 29, 89: 30

95: 4

96: 1, 96: 2, 96: 3, 96: 4, 96: 5

97: 1, 97: 2, 97: 3, 97: 4, 97: 5

99: 1, 99: 2, 99: 3, 99: 6, 99: 7, 99: 8

104: 5, 104: 6, 104: 7, 104: 8, 104: 9

111: 1, 111: 2, 111: 3

Bibliography

Alison Ross: http://news.bbc.co.uk/2/hi/science/nature/4260334.stm; and "Tracking Y Chromosomes Through Time", from the magazine,*Natuurwetenschap en techniek,* Oktober 2009 ,cf a blog: http://geographer-at-large.blogspot.in/2012/02/tracking-y-chromosomes-through-time.htm

Arrhenius, S., *Worlds in the Making: The Evolution of the Universe*. New York, Harper & Row, 1908.

Behrensmeyer, A. K., Damikh, J. D. et al., *Terrestrial ecosystems through time*, Chicago Univerisity Press, Chicago, I'll., 1992.

Bernal, J.D., *Physical basis of Life*, Routledge and Paul, London, 1951.

Bernal, J. D., *The Origin of Life*, World Pub. Co., Cleveland, 1967.

Bruce Alberts, Dennis Bray, et al., *Essential Cell Biology*, Garland Science, 2013.

Byrne, R., *The thinking ape*, Oxford University Press, 1995.

Cairns-Smith, A. G. and Hyman Hartman, *Clay minerals and the origin of life*, Cambridge University Press, 1986.

Calvin, M., *Chemical Evolution*, Oxford U. Press, New York, 1969.

Cann, R. L., Stoneking, M., & Wilson, A. C., "Mitochondrial DNA and human evolution", *Nature*, 325, (6099): 31–36, 1987, Cf.

http://en.m.wikipedia.org/wiki/Recent_African_origin_of_modern_humans#Mitochondrial_DNA

Carroll, R.L., *Vertebrate Paleontology and Evolution*. W.H. Freeman and Company, New York, 1988.

Carroll, Robert L., The origin and early radiation of terrestrial vertebrates, *J. of Paleontology*, Vol.75: 6, pp. 1202-1213, 2001.

Charles Darwin, *The Origin of Species* (with introduction by Julian Huxley, Signet Classics, 2003.

Colin A. Ronan, *Deep Space*, Crescent, 1987.

Condie Kent, C, *Plate Tectonics*, Butterworth-Heinemann; 4 edition, 1997.

Convey Morris, S., *The Crucible of Creation: The Burgess Shale and the Rise of Animals*, Oxford University Press, 1998.

Cook, Robert B., Man and the Biogeochemical Cycles: Interacting with the Elements, *Environment: Science and Policy for Sustainable Development*, Vol. 26, Issue 7, pp. 10-40,

Cowan, R., *History of life*, Blackwell Scientific Publications, Oxford, 1989.

Cuny, Hilaire, *Louis Pasteur: the man and his theories*, Translated by P. Evans. London: The Scientific Book Club, 1965.

Dawkins, R., *River out of Eden. A Darwinian view of life*, Weidenfeld & Nicholson, London, 1989.

Dubois, Rene J., *Louis Pasteur Free Lance of Science*, BiblioLife, 2015.

Dronamraju, K. R. (editor), *Haldane and Modern Biology*, Johns Hopkins University Press, Baltimore, Maryland, 1968.

Ehrlich, P.R., and Ehrlich, A.H., *Population Resources environment*, W. H. Freeman, New York, 1972.

Eldredge, N.; Gould, S.J., Punctuated Equilibria: an Alternative to Phyletic Gradualism. In Schopf, J. M. (ed.), *Models in Palaeobiology*. Freeman, Cooper & Co, 1972.

Farley, John, *The spontaneous generation controversy from Descartes to Oparin*.Johns Hopkins University Press, Baltimore, 1977.

Francesco Redi, *Experiments On the Generation of Insects*, Paperback reprint, Ulan Press, 2012

Fred Hoyle, "Hoyle on Evolution," *Nature*, Vol. 294, November 12, 1981.

Fox, Douglas, "Primordial Soup's On: Scientists Repeat Evolution's Most Famous Experiment", *Scientific American*, History of Science (Scientific American Inc.), 2007.

Fox, Sidney W., *The origins of prebiological systems and of their molecular matrices*, Academic Press, New York, 1965.

Fox, Sidney W., "The Protein Theory of the Origin of Life," *American Biology Teacher,* Vol. 36, pp. 161-172, 1974.

Gould,S. J., *The wonderful life*, Penguin, Harmondsworth, 1991.

Gould, S. J., *Ontogeny and Phylogeny*. Cambridge, MA: Harvard University Press, Belknap Press, 1977.

Gould Stephen, J and Niles Eldredge, Punctuated Equilibrium: The Tempo and Mode of Evolution Reconsidered," *Paleobiology*, 3 (2), pp. 115-151, 1977.

Haeckel, Ernst. *The Evolution of Man: A Popular Exposition of the Principal Points of Human Ontogeny and Phylogeny*, 1866; New York, 1896.

Haldane, J.B.S., *Origin of Life*, Rationalist Annual, 1929.

Harold, Franklin M., *In Search of Cell History: The Evolution of Life's Building Blocks*, University of Chicago Press, Chicago, p. 164, 2014.

Harun Yahya, *Darwinism Refuted*, Goodword Books, New Delhi, 2003

Harun Yahya, *The Miracle of Human Creation*, Goodword Books, New Delhi, 2003.

Harvey, William, History.com.
http://www.history.com/encyclopedia.do?articleId=211492
(Accessed November 18, 2009).

- See more at: https://embryo.asu.edu/pages/william-harvey-1578-1657#.dpuf

Helge Kragh, *Cosmology*, Princeton, NJ, USA: Princeton University Press, Princeton, NJ, 1996.

Helge Kragh, Cosmology and Controversy: *The Historical Development of Two Theories of the Universe*, Paper back, Princeton University Press, 1999.

Hubble, E. P., "A relation between distance and radial velocity among extra-galactic nebulae", *Proc. Natl. Acad. Sci.*, USA, 15(3): 168-73, 1929.

Ibrahim, I. A., *A brief Illustrated guide to understanding Islam*, Dar-us-Salam Publications, 2nd edition, Riyadh, 1999.

John S. Wilkins, "Spontaneous Generation and the Origin of Life", http://www.talkorigins.org/faqs/abioprob/spontaneous-generation.html

Keith L. Moore, T.N. Persaud, and Mark G. Torchia, *The Developing Human*: Clinically Oriented Embryology, Saunders; 2011

Keith S. Thomson,"Ontogeny and Phylogeny Recapitulated," *American Scientist*, Vol. 76, May/June 1988.

Kemp, T. S., *Fossils and Evolution*, Oxford University Press, 1999.

Leakey, Mary D., "Discoveries at Laetoli in Northern Tanzania", *Proceedings of the Geologists' Association*. 92 (2), pp. 81–86. 1981.

Lemaître, G., "Expansion of the universe, The expanding universe", *Monthly Notices of the Royal Astronomical Society*, Vol. 91, 490-501, 1931.

Lennox, James G., Aristotle's philosophy of biology: studies in the origins of life science. Cambridge, UK; New York: Cambridge University Press, 2001.

Lynn Margulis, *Origin of Eukaryotic Cells*, Yale University Press, 1971.

MailOnline: Early Human Migration, 2011: http://www.dailymail.co.uk/sciencetech/article-1351322/Man-left-Africa-Arabian-Peninsula-65-000-years-earlier-previously-thought.html

Margulis, L., Ed. *Origins of Life I*, Gordon and Breach, New York, 1970.

Margulis, L., Ed. *Origins of Life II*, Gordon and Breach, New York, 1971.

Margulis, L. and Schwartz K. V., *Five kingdoms, an illustrated guide to the phyla of life on Earth*, W. H. Freeman, New York, 1998.

Maurice Bucaille, *The Bible, the Qu'ran and Science: The Holy Scriptures Examined in the Light of Modern Knowledge* (Translated from French by Alastair D. Panel), CreateSpace Independent Publishing Platform, 2003.

Mc Neely, J.d. et al., *Conserving the Worlds Biological Diversity*; International Union for the Conservation of Natural Resources, Gland, Switzerland, 1990.

Meyer, A. W., *The Rise of Embryology*, Stanford University Press, California , 1939.

M. Gallego Llorente et al., Ancient Ethiopian genome reveals extensive Eurasian admixture throughout the African continent, *Science* 2879 Published online 8 October 2015 or http://www.sciencemag.org/content/early/2015/10/07/science.aad2879

Miller. Stanley L., "Production of Amino Acids under Possible Primitive Earth Conditions," *Science*, Vol. 117, pp. 528-529, 1953.

Miller, S. L. (1992) "The prebiotic synthesis of organic compounds as a step toward the origin of life", in *Major Events in the History of Life* (ed. J. W. Schopf). Boston: Jones and Bartlett, pp. 1–28.

Miller, S. L., J. W. Schopf, and A. Lazcano, Oparin's "Origin of Life: sixty years later". *Journal of Molecular Evolution* 44 (4):3513, 1997.

Miller, S. L. and L. E. Orgel, The *Origins of Life on the Earth,* Prentice-Hall, Englewood Cliffs, New Jersey, 1973.

Miller, Stanley L., Harold C. Urey, "Organic Compound Synthesis on the Primitive Earth", *Science* 130 (3370): pp. 245–51, 1959.

Morris, S. Convey, *The Crucible of Creation: The Burgess Shale and the Rise of Animals*, Paperback , Oxford University Press, 1999.

Naseem Ahmad, *Science in Islam*, Reference Press, New Delhi, 2003.

Neil Campbell and Reece, J., *Biology*: 7th Edition, Pearson, Benjamin, Cummings, 2005.

Oparin, A.I., *The Origin of Life*, (translated by Sergius Morgulis), The Macmillan Co., New York, 1938.

Oparin, A. I., The Origin of Life on the Earth (translated by Ann Synge), Academic Press Inc., New York, 1957.

Orgel Leslie E., *The Origins of Life*, Paperback, John Wiley & Sons Inc., 1973.

Orgel Leslie E., "The Origin of Life on the Earth", *Scientific American*, Vol. 271, October 1994.

Orgel, Leslie, "The origin of life -- a review of facts and speculations," *Trends in Biochemical Sciences,* 23, pp. 491-495, Dec.1998.

Orgel, Leslie E., "Some consequences of the RNA world hypothesis". *Origins of Life and Evolution of the Biosphere.* (Kluwer Academic Publishers), 33 (2): pp.211–218, 2003.

Richardson, M. K., J. Hanken, L. Selwood, G. M. Wright, R. J. Richards, C. Pieau, and A. Raynaud. "Haeckel, Embryos, and Evolution." *Science* 280 (1997):983-984.

Robert Shapiro, Origins: *A Skeptic's Guide to the Creation of Life on Earth*, Summer Books, New York, 1986.

Sabiha Saadat, “Human Embryology and the Holy Quran: An Overview,” *Int.J.Health Sci. (Qassim)*, 2009 Jan. 3(1), pp. 103-109.

Salisbury, Frank B., "Doubts about the Modern Synthetic Theory of Evolution," *American Biology Teacher*, 33 (6), pp. 335-338, 1971.

Shopf, J.W., *Cradle of life*, Princeton Univerisity Press, Princeton, NJ, 1999.

Simpson, G.G., Beck, W., An Introduction to Biology, Harcourt Brace and World, New York, 1965.

Singer, Charles Joseph, A history of biology to about the year 1900: A general introduction to the study of living things. 3rd ed. London: Abelard-Schuman, 1959.

Southwick, C.H., Ecology and the Quality of Our Environment, Von Nostrand Reinhold Company, New York, 1972

Southwick, C.H., *Global Ecology*, Sinauer Associates, Inc., Sunderland, Massachusetts, 1985.

Southwood, T.R.E., *The Story of Life*, Oxford University Press Inc., New York, 2003.

Stanley, S.M., *Earth and life through time*. W. H. Freeman. New York, 1989.

Stephen J. Gould, Wonderful Life - The Burgess Shale and the Nature of History, Paperback, W. W. Norton & Company, 1991.

Stork, Nigel E., "How many species are there?", *Biodiversity and Conservation,* 2 (3): pp. 215–232, 1993.

Taylor, D. J., G. W. Stout, N. P. O. Green, R. Soper, *Biological Science*, Cambridge University Press, Cambridge, 2008.

Tim Bromage, "Faces from the Past," *New Scientist*, Vol 133, issue 1803, p. 41, 1992.

Tudge, C., *The variety of life*, Oxford University Press, 2000.

Vigilant L, Stoneking M, Harpending H, Hawkes K, Wilson AC; Stoneking; Harpending; Hawkes; Wilson, "African populations and the evolution of human mitochondrial DNA", Science, 253 (5027): 1503–7, 1991, Cf. http://en.m.wikipedia.org/wiki/Recent_African_origin_of_modern_humans#Mitochondrial_DNA

Wächtershäuser, G. "Evolution of the First metabolic cycles", *Proceedings of the National Academy of Sciences of the United States of America.* 87 (1): 200–4, January 1990.

Wegener, Alfred, translated by John Biram, *The Origin of Continents and Oceans*, Dover Publications Inc., New York, 2012.

Wells, H.G.,*The Outline of History: A Plain History of Life and Mankind* (Mindvessel Edition), Figure 56: http://www.ibiblio.org/pub/docs/books/sherwood/Wells-Outline/Images/0056img.htm

Whittaker, R. H. and Likens, G.E., "Primary production: The biosphere and man", *Human Ecology*, Vol. 1, Issue 4, pp. 357-369, September 1973.

Whitaker, R. H. and Woodwell, G.M., "Measurement of net primary productivity of forests", pp. 159-75, In Duvigneaud, P., editor, *Productivity of Forest Ecosystems,* UNESCO, Paris, 1971.

Wilkins, John S., Spontaneous Generation and the Origin of Life, The Talk.Origins Archive, April 2004,

Wilson, R.C.L., Drury, S.A., and Chapman, J.l., *The great ice age. Climate change and life*, Routledge, London, 2000.

Zakir Naik, *The Quraan and the Modern Science,* Riyadh. 2007, pp. 60-74